Cognitive Intelligence and Robotics

Cognitive Intelligence refers to the natural intelligence of humans and animals, it is considered that the brain performs intelligent activities. While establishing a hard boundary that distinguishes intelligent activities from others remains controversial, most common behaviors and activities of living organisms that cannot be fully synthesized using artificial means are regarded as intelligent. Thus the acts of sensing and perception, understanding the environment, and voluntary control of muscles, which can be performed by lower-level mammals, are indeed intelligent. Besides the above, advanced mammals can perform more sophisticated cognitive tasks, including logical reasoning, learning, recognition, and complex planning and coordination, none of which can yet be realized artificially to the level of a baby, and thus are regarded as cognitively intelligent.

This book series covers two important aspects of brain science. First, it attempts to uncover the mystery behind the biological basis of cognition, with a special emphasis on the decoding of stimulated brain signals or images. Topics in this area include the neural basis of sensory perception, motor control, sensory-motor coordination, and understanding the biological basis of higher-level cognition, including memory, learning, reasoning, and complex planning. The second objective of the series is to publish consolidated research on brain-inspired models of learning, perception, memory, and coordination, including results that can be realized on robots, enabling them to mimic the cognitive activities performed by living creatures. These brain-inspired models of machine intelligence complement the behavioral counterparts studied in traditional artificial intelligence.

The series publishes textbooks, monographs, and contributed volumes.

Bin-Bin Hu · Hai-Tao Zhang

Multi-Robot Systems

Coordinated Fencing Control and Applications

Springer

Bin-Bin Hu
Faculty of Science and Engineering
University of Groningen
Groningen, The Netherlands

Hai-Tao Zhang
School of Artificial Intelligence
and Automation
Huazhong University of Science
and Technology
Wuhan, Hubei, China

ISSN 2520-1956 ISSN 2520-1964 (electronic)
Cognitive Intelligence and Robotics
ISBN 978-981-97-9213-9 ISBN 978-981-97-9214-6 (eBook)
https://doi.org/10.1007/978-981-97-9214-6

This Springer imprint is published by the registered company Springer Nature Singapore Pte Ltd.
The registered company address is: 152 Beach Road, #21-01/04 Gateway East, Singapore 189721, Singapore

Preface

Over the decades, significant advances have been made in the field of multi-robot system research, encompassing various applications such as unmanned aerial vehicles (UAVs), unmanned surface vehicles (USVs), and autonomous ground vehicles (AGVs). Compared to a single robot, multi-robot coordination offers higher efficiency, greater robustness, broader operational ranges, and improved resilience, making it increasingly vital in various industrial and civil applications. One of the interesting problems in multi-robot coordination is multi-robot coordinated fencing, which governs a team of robots to form a protective formation around a target, has drawn significant attention, and proved to be useful in practice such as strategic defense and cargo transportation. However, real-world applications often involve complex target characteristics, such as varying dynamics and multiple targets, which can lead to failures in maintaining the fence missions. Additionally, due to sensor costs and environmental constraints, some robots may only have access to directional constraint information, which further intensifies challenges to the coordinated fencing.

To this end, this monograph systematically introduces the current theoretical and practical progress for designing coordinated multi-robot fencing control algorithms and their applications to unmanned vehicles, which can help the readers have an overview and understand the research lines and future directions of coordinated fencing. The readers can benefit from learning how to formulate the practical fencing mission into a mathematical problem, and how to design the theoretical controller for practical applications with rigorous guarantees. Overall, the monograph is divided into three parts.

Firstly (Chap. 1), we introduce the background of multi-robot systems and USV swarms. We then dive into related works on coordinated fencing control, categorizing them into two research directions, namely trajectory fencing and convex-hull fencing. Along these research lines, we present the necessary tools for the convergence analysis later. Finally, we outline the structure of the entire monograph.

In the first part (Chaps. 2–5), we introduce the basic fencing control via neighboring angle repulsion for a constant-velocity target in Chap. 2, which lays a groundwork for investigating complex coordinated fencing in the following chapters. Next, we design a coordinated controller using input-to-state stability in Chap. 3, achieving complex fencing with an evenly rotating formation. In Chap. 4, we leverage the output regulation theory to tackle a more complex fencing problem with a varying-velocity target. Finally, in Chap. 5, we propose a formal long-term task execution framework, which can handle the general fencing missions for a target in both 2D and 3D.

In the second part (Chaps. 6–7), we address the multiple-target scenario. Specifically, we propose an adaptive fencing controller to govern a team of robots to form a large circle, ensuring that the smaller circle formed by the targets is fenced by the robot group in Chap. 6. Then, in Chap. 7, we consider a more challenging scenario where the targets can be scattered, where we propose a flexible subgrouping and ordering methodology to fence the scattered targets in challenging environments flexibly.

In the third part (Chap. 8), we introduce a bearing-only fencing controller to accomplish the coordinated target-fencing mission. Unlike the previous chapters (i.e., Chaps. 2–7) relying on the relative position or distance between the robot and target, such a bearing-only approach only requires comparatively inexpensive sensors, such as pinhole cameras and wireless sensor arrays. While this method can be applied to large-scale systems, it inevitably induces more challenging issues due to bounded bearing errors.

Finally (Chap. 9), we summarize the entire monograph and outline potential future directions.

As a professional reference monograph, this book covers various cooperative fencing research topics, benefiting readers with varying levels of expertise in broad fields such as multi-robot cooperation, fencing control theory, unmanned vehicle formation, and control engineering. It is an invaluable resource for professional researchers, graduate students, university faculty, and others. This book will help readers systematically study the related fencing theory and its applications to different unmanned vehicles.

This book consists of various works during my previous research period, as well as my supervisor, cooperators, and group members, with whom we have worked over the years and have made it possible. Special thanks would first go to my supervisor Hai-Tao Zhang for his detailed guidance. We are also grateful to Prof. Zhiyong Chen for his outstanding mathematical skills and rigorous requirements for my work. Thanks for the guidance of Prof. Jun Wang from Hong Kong City University. We are also grateful to all the group members of the Intelligent Manufacturing and Data Science Laboratory (IMDS), especially Haofei Meng, Bin Liu, Bowen Xu, Jianing Ding, Fulong Hu, Yaozhong Zheng, Yifan Hu, etc., for the constant feedback and support. Then, we thank Jingying Chen, and Niraja Deshmuk in the publication team at Springer for her assistance. Finally, sincere thanks to our beloved families for their support all the time.

Research efforts summarized in this monograph were supported by the National Natural Science Foundation of China under Grant 62225306; in part by the National Key R&D Program of China under Grant 2022ZD0119601; in part by the National Natural Science Foundation of China under Grant U2141235 and Grant 61803168.

Groningen, The Netherlands
Wuhan, China
September 2024

Bin-Bin Hu
Hai-Tao Zhang

Contents

1 Introduction to Coordinated Multi-robot Fencing Control 1
1.1 Background 1
1.2 Related Works 4
1.3 Preliminary 7
1.4 Book Outline 9
References 11

Part I Single-Target Scenario

2 Distributed Fencing Control via Neighboring Angle Repulsion 17
2.1 Introduction 17
2.2 Problem Formulation of Fencing a Constant-Velocity Target 19
2.3 Theoretical Convergence Analysis 22
2.4 Simulation Results 30
2.5 Conclusion 32
References 32

3 Distributed Fencing Control via Input-to-State Stability 35
3.1 Introduction 35
3.2 Problem Formulation of Fencing with an Evenly Rotating Formation 37
3.3 Distributed Target Estimation Under Intermittent Topologies 41
3.4 Controller Design and Convergence Analysis 42
3.4.1 Target Fencing Control—Stage 1 42
3.4.2 Equal Fencing with Evenly Rotating Formation—Stage 2 44
3.5 Algorithm Verification 48
3.6 Conclusion 53
References 53

4 Distributed Fencing Control via Output Regulation Theory 55
4.1 Introduction 55
4.2 Problem Formulation of Fencing a Varying-Velocity Target 57
4.3 Internal-Model Controller Design and Convergence Analysis 59
4.4 Rigid Fencing Pattern and Analysis 71
4.5 Simulation Results 74
4.6 Conclusion 79
References 79

5 Distributed Fencing Control via Long-Term Task Execution 81
5.1 Introduction 81
5.2 General Fencing Problem Formulation 83
5.2.1 Long-Term Task Execution 83
5.2.2 Ordering-Flexible Target Fencing 86
5.2.3 Encoding Long-Term Target Fencing 88
5.3 Long-Term Task Execution Controller Design and Convergence Analysis 92
5.4 Algorithm Verification 103
5.4.1 2D Experiments 103
5.4.2 3D Simulations 108
5.5 Conclusion 112
References 113

Part II Multiple-Target Scenario

6 Coordinated Multiple-Target Fencing Control and Collision Avoidance 117
6.1 Introduction 117
6.2 Problem Formulation of Fencing Multiple Varying-Velocity Targets 119
6.3 Estimation of the Center of Targets and Analysis 122
6.4 Equal-Distance Fencing and Convergence Analysis 125
6.5 Equal-Distance Fencing with Collision Avoidance and Convergence Analysis 130
6.6 Algorithm Verification 132
6.6.1 A Multi-USV Experiments Platform 132
6.6.2 Setup and Results 133
6.7 Conclusion 135
References 137

7 Coordinated Multiple-Scattered-Targets Fencing via Flexible Subgrouping and Ordering 139
7.1 Introduction 139
7.2 Preliminaries 141
7.2.1 Task Set 141
7.2.2 Multi-robot Task Execution 142
7.2.3 Multiple Scattered Targets 143

7.3 Problem Formulation of Fencing Multiple Scattered Targets 144
7.3.1 Flexible Subgrouping for MSTs 144
7.3.2 Flexible-Ordering Fencing for Each Specific Target 144
7.4 FSOC Optimization 146
7.5 Convergence Analysis 149
7.5.1 Subgrouping Convergence 150
7.5.2 Fencing Convergence 151
7.6 Simulation Results 159
7.7 Conclusion 164
References 168

Part III Bearing-Only Scenario

8 Bearing-Only Fencing via Persistent-Exciting Condition 173
8.1 Introduction 173
8.2 Problem Formulation of Bearing-Only Target Fencing 175
8.3 Perturbed-Stable Controller and Convergence Analysis 180
8.3.1 Target Fencing Control 180
8.3.2 Vessel Regulation Control 187
8.4 Algorithm Verification 189
8.5 Conclusion 193
References 193

9 Summary and Future Work 197
9.1 Summary 197
9.2 Future Work 198

Acronyms

2D	Two-Dimensional
3D	Three-Dimensional
AG	Accelerator Gyroscope
AGV	Autonomous Ground Vehicles
AMR	Autonomous Mobile Robot
BCS	Base Control Station
CBF	Control Barrier Functions
CLF	Control Lyapunov Functions
CS	Control Station
DC	Direct Current
DGPS	Differential Global Positioning System
FN	Front Neighbor
FSOC	Flexible Subgrouping and Ordering Coordination
GPS	Global Positioning System
ISS	Input-to-State Stability
KKT	Karush–Kuhn–Tucker
LAN	Local Area Network
LED	Light Emitting Diode
LTTE	Long-Term Task Execution
MAS	Multi-Agent System
MCS	Motion Capture System
MCU	Microcontroller Unit
MRS	Multi-Robot System
MST	Multiple Scattered Targets
NID	Near-Identity Diffeomorphism
NN	Neural Network
PE	Persistently Exciting
QP	Quadratic Programming
RN	Rear Neighbor
RNN	Recurrent Neural Network
ROS	Robot Operating System

SD	Secure Digital
SLAM	Simultaneous Localization and Mapping
TCP	Transmission Control Protocol
UAV	Unmanned Aerial Vehicle
UGV	Unmanned Ground Vehicle
USV	Unmanned Surface Vehicle

Chapter 1
Introduction to Coordinated Multi-robot Fencing Control

1.1 Background

In the past decades, the coordination of multi-robot systems has attracted a great deal of attention due to its low cost, high efficiency, large range, and strong robustness [1]. Technically speaking, the fundamental theory of the coordination of multi-robot systems is swarm intelligence, which was initially inspired by observing the common characteristics of group behaviors in nature to reveal their underlying mechanisms and principles. As shown in Fig. 1.1, complex behaviors such as the coordinated flight of bird flocks during migration, collective evasion of fish schools in the face of threats, collaborative transport by ant colonies during foraging, and cooperative work of bee swarms in nest building have all inspired swarm intelligence research. As for the word "swarm intelligence", it originates from [2], where the scientist Gerardo first discussed how the behaviors of swarms and other social insects inspired robotic control strategies based on multi-agent systems. Then, Eric further introduced the modeling and algorithm design of swarm intelligence and the potential and advantages of applying swarm intelligence algorithms in artificial systems, including robotics, distributed sensor networks, and optimization problems [3].

Initially, the research on swarm intelligence mainly focused on the theoretical mechanism modeling of the swarm behaviors, which involves approximating the rapid aggregation and dispersion of animal groups using network rules or differential mathematical equations. As early as 1986, the computer scientist Reynolds et al. first proposed the three basic rules of swarm groups: "repulsion", "aggregation", and "speed matching", and reproduced this groundbreaking swarm behavior through computers [4], which has also become a classic rule widely accepted and adopted by many scholars. Later, Vicsek et al. introduced the Vicsek model for self-driven

B.-B. Hu and H.-T. Zhang, *Multi-Robot Systems*, Cognitive Intelligence and Robotics,
https://doi.org/10.1007/978-981-97-9214-6_1

Fig. 1.1 Swarm behaviors in the natural world. **a** Bird flocking. **b** Fish schooling. **c** Ant colony. **d** Bee swarm

particle systems to study the behavior and phase transitions of particle groups [5, 6]. Couzin et al. further proposed the Couzin model, which considers three interaction rules: repulsion, alignment, and attraction, simulating the effects of limited visual perception [7]. Then, Andrea and Giorgio et al. applied the spin model to the study of starling flocks, revealing the relationship between the group dynamics and spin models [8]. With the development of robotics, the research on swarm intelligence has gradually moved toward experimental validation, which attempts to simulate and replicate certain behaviors of biological swarms using actual robots. For instance, Radhika et al. [9] from Harvard University studied a large-scale self-organizing robotic system, where they used small, affordable 2D robots called Kilobots for communication and coordinated formation to mimic the complex swarm behaviors in nature. Additionally, Tamas et al. [10] used drone swarms to simulate swarm behaviors in constrained environments, while Xing Zhou et al. considered the swarm behaviors of drones in more complex, dense environments [11]. Furthermore, Zhang et al. implemented the swarm cooperative control strategies on water surfaces using multi-unmanned-surface vehicle (USV) systems.

Meanwhile, due to the increasingly complex tasks, the coordination of unmanned vehicles has attracted a great deal of attention. For instance, as a new type of intelligent maritime platform, unmanned surface vessels (USVs) can possess advantages such as recyclability, low cost, strong maneuverability, and long-term operation [12]. In this

Fig. 1.2 Different application areas of our self-established USVs. **a** HUSTER-12 USV: River exploration. **b** HUSTER-16 USV: Seafloor mapping. **c** HUSTER-30 USV: Hydrographic surveying. **d** HUSTER-68 USV: Ocean surveying

way, such kinds of USVs have gradually played an important role in various applications such as intelligence surveillance, precision strikes, resource exploration, and water quality monitoring and regulation [13–16]. As shown in Fig. 1.2, we illustrate different types of our self-established USVs. Due to the limited payload capacity of a single USV, a natural idea is to develop the coordination of USVs, which can offer advantages such as high efficiency, wide coverage, large scale, robustness, and strong mobility, and boost tremendous potential for future applications. From the perspective of cutting-edge theory, the journal Science's publication titled "The Power of the Crowd" [17] also emphasizes the need to combine collective intelligence with machine performance to address rapidly growing human challenges. In summary, the research of combining swarm control technologies with USV swarm will contribute to the further development and utilization of marine resources, which can effectively enhance efficiency and coverage in target protection. However, in various practical applications, there exist some complex target characteristics such as varying dynamics and multiple targets, which can lead to failures of coordinated fencing missions. Moreover, due to sensor costs and environmental constraints, USVs may only obtain directional constraint information, posing challenges to coordinated fencing.

To this end, it is an urgent mission to address the aforementioned challenging issues of complex target dynamics and sensor constraints and incorporate the control algorithm design with practical applications for coordinated fencing control of unmanned vehicles.

1.2 Related Works

In this section, we will systematically present the research line of the previous coordinated fencing works.

First of all, the theoretical support for the coordinated fencing of multi-robot systems originates from the multi-agent control theory [18–20], which has been extensively studied in the past decades. With the continuous emergence of research hotspots, an increasing number of scholars have devoted themselves to relevant research in this field, which has further promoted the rapid development of multi-agent theory and swarm theory [21, 22], such as unmanned systems [23–25], distributed sensor networks [26–28], mobile industrial robots [29–31], and smart grid [32–34]. Among them, coordinated formation, as one of the most active research areas in recent years, has been fully explored and a large number of research works have emerged. According to different methodology mechanisms, the research directions of collaborative formation include leaderless formation method [35, 36] and leader–follower formation method [37, 38], respectively.

In leaderless formation, the group of agents does not predefine an agent as the "leader", whereas all the agents mainly form and track the desired formation through interactions between neighboring agents [39, 40]. This greatly facilitates the redundancy and fault-tolerant design of the controller, making the formation system more robust in practice. However, in leader–follower formation, agents are usually divided into two groups, namely the leader group and the follower group, respectively. In such kind of a framework, in addition to interacting with the neighboring follower agents, the follower agent also needs to interact with the leader agent to ultimately achieve the desired formation [41, 42].

When the coordinated formation becomes more complex, the leader–follower formation algorithm has evolved many interesting new branches, among which the containment control problem [43] and the fencing control problem [44] are two typical representatives. Generally speaking, containment control mainly involves cooperatively controlling the follower agents to be contained by the convex hull formed by the leader agents whereas keeping maneuvering. Essentially, the core problem is still similar to the consensus problem, where the main application scenarios include hazardous materials handling, search and rescue, and collaborative transportation [45]. For instance, a group of heterogeneous robots needs to move from one area to another, but only a small number of them are equipped with advanced detection equipment that can detect dangerous obstacles around them. Usually, the small number of robots is defined as the leader group, and the rest of the robots are defined as the follower group. Then, by detecting obstacles, the leader group can form a safe moving area

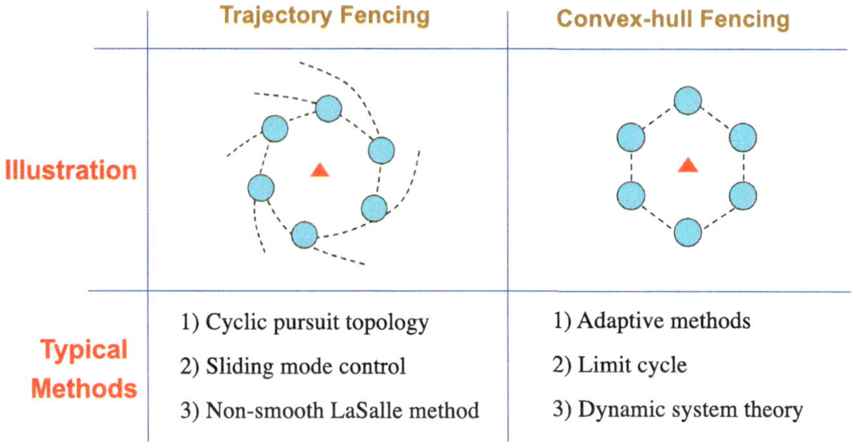

Fig. 1.3 Illustration and representative methods of the trajectory and convex-hull fencing

(convex hull), and the containment controller will govern the follower robot group to enter the safe area and move with it. Finally, the robot groups can safely arrive at another specified area. In recent years, containment control has been widely studied, mainly including first-order systems [46], second-order systems [47], complex unmanned systems, etc. [48].

In contrast, the corresponding fencing control is to collaboratively govern multiple follower agents to fence and protect the leader agent(s) [44], which can be regarded as an inverse problem of containment control. Compared with the containment control, the fencing control is more closely integrated with the unmanned vehicle swarm and has a wider range of practical usage. According to different fencing strategies, the fencing control is approximately divided into trajectory-fencing and convex-hull fencing methods, where some illustrative examples are given in Fig. 1.3. To proceed, we will introduce the related works of these two methodologies in the following part.

(i) Trajectory-Fencing Scenario: For the trajectory-fencing control algorithms, the core idea is to drive a group of agents to encircle a target using rotating trajectories [49]. In recent years, there has been extensive research on trajectory-fencing algorithms. Marshall et al. [49] first proposed a formation algorithm based on the cyclic pursuit principle to fence a stationary target. Kim et al. [50] proposed a 3D distributed cyclic pursuit controller to form a predefined tracking formation and fence a moving target. Guo et al. [51] proposed a distributed motion target-fencing method based on relative neighboring positions. Zheng et al. [52] designed a bearing-only controller, which is capable of fencing both point targets and disk-shaped targets. However, the above works [49–52] assume that target information is available to all agents, which is limited by the theoretical bottlenecks. Subsequently, many scholars have further investigated a more practical scenario where target information is partially available to some agents. In this endevator, Sharma et al. [53] proposed a robust fencing controller based on partially received target information. Yu et al. explored various scenarios including fixed directed topologies [54], switching topologies [55], and limited individual target velocities [56], and designed cooperative fencing controllers for the affine systems. Shi et al. [57] studied scenarios where information about multiple stationary targets is partially known. Dou et al. [59] considered the moving target estimation and collision avoidance, and proposed a cooperative fencing strategy based on output regulation theory. To reduce costs and better match practical conditions, Dou et al. [58] investigated a bearing-only distributed target localization and designed cooperative fencing strategies for unknown targets. Yang et al. [60] further studied the single target-fencing formation under more complex conditions.

However, due to practical application limits, the aforementioned fencing works [49–60] still focused on simple first-order or second-order linear systems and have not yet been integrated with complex real-world systems. In recent years, a series of fencing formation works with unmanned vehicles have emerged. For instance, Dong et al. [61, 62] studied the time-varying formation strategies for multiple UAVs cooperating in fencing tasks. Peng et al. [63, 64] investigated target-fencing strategies with multi-USV systems. It is still worth mentioning that the existing works only focused on a single USV fencing a single moving target, whereas the multi-USV fencing problem remains an open problem.

(ii) Convex-Hull Fencing Scenario: Despite the previous trajectory-fencing methods employing agents' motion trajectories to fence a target, they cannot rigorously guarantee that the target is continuously fenced all along from a mathematical view. Therefore, it motivates another direction of the convex-hull fencing methods. Unlike trajectory fencing which relies on trajectories, the convex-hull strategy strictly fences a target at every moment using the convex hull formed by all agents. Moreover, the convex-hull fencing algorithms can be further categorized into basic and complex fencing according to their missions and the complexity of the target dynamics.

The basic fencing tasks typically involve a straightforward scenario of the fence tasks with simple target dynamics like stationary or uniformly moving targets. These tasks have strong engineering backgrounds and relatively straightforward practical applications, leading to numerous research advancements in recent years. Liu et al. [65] proposed a cooperative hierarchical fencing strategy based on output regulation theory, which achieves the cooperative encirclement of a static target. Furthermore, Chen [66] introduced a new cooperative fencing controller incorporating attraction, repulsion, and rotation, which is capable of fencing a stationary target without a predefined radius. Inspired by this work, Kou et al. [67] extended these concepts to moving targets, designing an integrated estimation-control structure to achieve fencing. Notably, the cooperative formations exhibited interesting patterns of translation and rotation depending on whether individual repulsion functions were added to the estimator or not. The work by Kou et al. [67], focusing on first-order linear systems, is relatively uncommon in practical engineering. Subsequently, this research was extended to encircling moving targets with second-order linear multi-agent systems, yielding similar conclusions [68].

However, since the basic fencing strategies are only limited to simple tasks that cannot deal with dynamic and complex environments, more researchers have turned their focus toward complex fencing tasks, such as complex formation patterns, a varying-velocity target, and multiple targets, which are more suitable for environments with diverse and challenging dynamics. Wang et al. [69] designed a fencing strategy based on limit cycles to fence moving targets, capable of forming circular formations. Xu et al. [70] studied multi-agent fencing strategies for moving targets under input saturation conditions. Chen et al. [44] proposed an integrated estimation-control framework using connected topologies to fence multiple stationary targets with multiple agents. Shi et al. [71] further considered multi-target fencing with second-order linear systems. To align with practical engineering needs, Shoja et al. [72] extended these concepts to nonlinear and non-identical multi-agent systems, devising adaptive control strategies accordingly. Later, Li et al. [73] introduced a continuous multi-target estimator and designed a corresponding fencing strategy. For even more complex dynamic systems, Li et al. [74] developed cooperative fencing controllers based on Lagrangian multi-agent systems, ensuring collision avoidance among agents.

In summary, among the aforementioned research works, numerous scholars globally have conducted extensive theoretical studies on cooperative fencing control problems, where the convex-hull cooperative fencing control algorithm stands out for its theoretical ability to guarantee the effectiveness and success rate of fencing in practice. However, due to its relatively late inception, many related studies are primarily focused on the basic fencing involving first-order and second-order linear systems, which have yet to integrate complex nonlinear system models in practical engineering scenarios. Furthermore, the issues of collision avoidance, complex motion target characteristics, formation control, and sensor limitation commonly encountered in practical engineering have not been fully addressed as well. Moreover, the majority of these studies remain in simulations and lack real-world experimental validation, thereby presenting challenges in demonstrating algorithm reliability. Despite some studies considering more practical and complex fencing tasks, they only focus on the characteristics of the targets being fenced, without addressing issues such as formation control of agents, multiple targets, and other practical aspects crucial in real-world applications.

Accordingly, the existing challenges regarding formation patterns, complex target behaviors, and sensor limitations still require further research, which motivates the publishment of this book to accelerate both theoretical innovations and practical applications of cooperative fencing algorithms in practice.

1.3 Preliminary

In this section, we introduce the graph theory, matrix theory, and stability theory for coordinated fencing in this monograph.

First of all, we define the notations throughout this monograph. The notations $\mathbb{R}$, $\mathbb{R}^+$ denote the real numbers and positive real numbers, respectively. The notation $\mathbb{R}^n$ represents the n-dimensional Euclidean space. The notations $\mathbb{Z}$ and $\mathbb{Z}_i^j$ represent the integer number and the integer set $\{m \in \mathbb{Z} \mid i \leq m \leq j\}$, respectively. The notation $\| * \|$ denotes the Euclidean norm of the vector $*$. The notation $*^{-1}$ represents the inverse of the square matrix $*$. The notation $*^{\mathsf{T}}$ denotes the transpose of the matrix $*$. The notation $\int *(s)d(s)$ represents the integral of the function $*$. The notation $\dot{*}$ represents the derivative of the function $*$ with respect to time t. The notations $* \succ 0$ and $* \prec 0$ represent the positive-definite and negative-definite matrices $*$, respectively. The identity matrix is denoted by $I_n := \mathrm{diag}\{1, 1, \ldots, 1\} \in \mathbb{R}^{n\times n}$, $(*_{ij})$ represents the (i, j)-th item of the matrix $*$. The Kronecker product is denoted by $\otimes$. The n-dimensional identity matrix is represented by I_n. The N-dimensional column vector consisting of all 1 is denoted by $\mathbf{1}_N$.

Lemma 1.1 ([18]) *For a connected and undirected topology $\mathcal{G}$, one has that*

$$\min_{x\neq 0, \mathbf{1}^{\mathsf{T}}x=0} \frac{x^{\mathsf{T}}Lx}{\|x\|^2} = \lambda_2(L),$$

where $x \in \mathbb{R}^n$, $L \in \mathbb{R}^{n\times n}$ is the corresponding Laplace matrix, and $\lambda_2 > 0$ represents the second smallest eigenvalue of the corresponding matrix L.

Lemma 1.2 (LaSalle's Invariance Principle [75, Theorem 4.4]) *Consider a system*

$$\dot{x} = f(t, x, u), \tag{1.1}$$

where $f : [0, \infty) \times \mathbb{R}^n \times \mathbb{R}^m \to \mathbb{R}^n$ is piecewise continuous in t and locally Lipschitz in x and u. The input $u(t)$ is a piecewise continuous, bounded function of t for all $t \geq 0$. Let $\Omega \in D$ be a compact set that is positively invariant with respect to (1.1)*. Let $V : D \to \mathbb{R}$ be a continuously differentiable function such that $\dot{V}(x) \leq 0$ in Ω. Define the set $\mathcal{E} := \{x \in \Omega : \dot{V}(x) = 0\}$, then every solution starting in Ω approaches M as $t \to \infty$.*

Lemma 1.3 (Input-to-State Stability [75, Theorem 4.19]) *Let $V : [0, \infty) \times \mathbb{R}^n \to \mathbb{R}$ be a continuously differentiable function such that*

$$\alpha_1(\|x\|) \leq V(t, x) \leq \alpha_2(\|x\|)$$
$$\frac{\partial V}{\partial t} + \frac{\partial V}{\partial x} f(t, x, u) \leq -W_3(x), \forall \|x\| \geq \rho(\|u\|) > 0,$$

$\forall (t, x, u) \in [0, \infty) \times \mathbb{R}^n \times \mathbb{R}^m$, where α_1, α_2 are the class $\mathcal{K}_\infty$ functions, ρ is a class $\mathcal{K}$ function, and $W_3(x)$ is a continuous positive-definite function on $\mathbb{R}^n$. Then, the system (1.1) *is input-to-state stable (ISS) with $\gamma = \alpha_1^{-1} \circ \alpha_2 \circ \rho$.*

Lemma 1.4 (Barbalat's Lemma [75, Lemma 8.2]) *Let $\alpha : \mathbb{R} \to \mathbb{R}$ be a uniformly continuous function on $[0, \infty)$. Suppose that $\lim_{t\to\infty} \int_0^t \phi(\tau)d\tau$ exists and finite. Then, $\phi(t) \to 0$ as $t \to \infty$.*

1.4 Book Outline

Focusing on the core scientific problem of distributed fencing control under different constraints, this monograph starts from the three perspectives of single-target fencing, multi-target fencing, and limited-sensing fencing. It explores various in-depth methodologies of the fencing theory in Fig. 1.4, where Chaps. 2–5 belong to Part I of a single-target scenario, Chaps. 6–7 belong to a multi-target scenario, and Chap. 8 belongs to the limited-sensing scenario. Moreover, extensive fencing experiments using multiple USVs and UAVs are conducted to validate the effectiveness of the algorithms in practice. In this chapter, we provide a necessary preliminary for the design of the fencing controller in the following chapters. The remainder of this monograph is organized below.

In Chap. 2, we propose a distributed fencing controller to govern a team of robots to fence a constant-velocity target cooperatively, which is a basic fencing problem. In particular, based on the robot–robot and robot–target distances, a cooperative fencing controller utilizing the nearest phase-angle repulsion is designed to accomplish three tasks: straight-line fencing, collision avoidance between individuals, and regular polygon formation. The convergence of the closed-loop system is demonstrated using the comparison principle and Barbalat's lemma. Finally, simulations are conducted to verify the effectiveness of the algorithm.

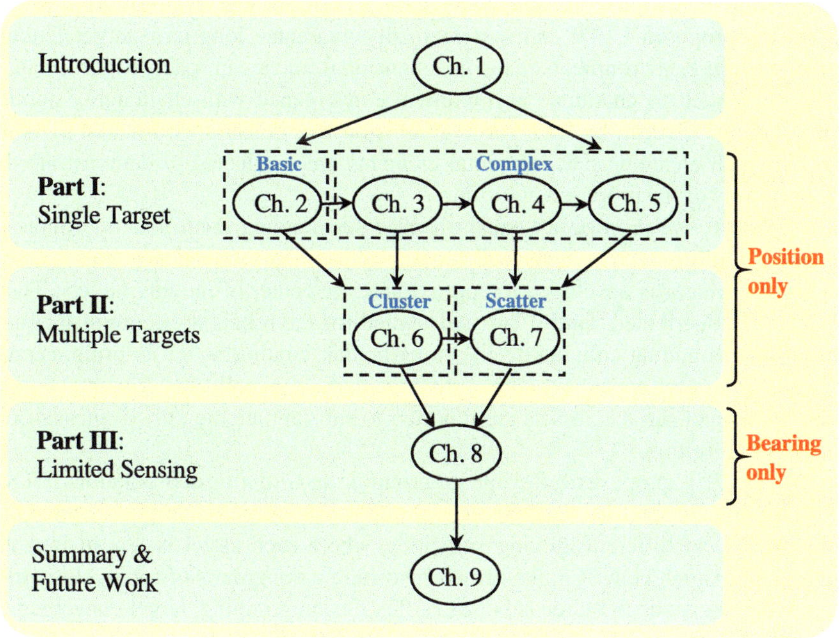

Fig. 1.4 The structure of the monograph and the relationship among different chapters

In Chap. 3, we propose a distributed fencing controller via input-to-state stability (ISS), which solves a complex fencing problem with an evenly rotating formation. Therein, considering the influence of intermittent communication topology and evenly rotating formations on the closed-loop system, an estimation-control hierarchical framework is established to decouple distributed target estimation and cooperative fencing missions. Then, a distributed fencing controller is designed via the ISS theorem, which achieves the rotating formation fencing for a moving target. Additionally, the convergence of the closed-loop system is guaranteed as well. Finally, simulations are carried out to verify the effectiveness of the algorithm.

In Chap. 4, we consider a more complex fencing problem with a varying-velocity target. Thereby, we propose a distributed fencing controller via output regulation theory, which cooperatively fences a moving target of variational velocity into a convex hull formed by the robots while maintaining a rigid formation. In particular, to attain a rigid formation with guaranteed collision avoidance, each controller consists of two terms: a dynamic regulator with an internal model to drive robots toward the moving target merely by position information feedback, and a repulsive force between each pair of adjacent robots. Significantly, rigorous analysis is provided to eliminate the strong nonlinear couplings induced by the label-free property. Finally, the effectiveness of the algorithm is substantiated by simulations.

In Chap. 5, we propose a distributed long-term task execution (LTTE) algorithm for the general ordering-flexible target-fencing problem. Particularly, by designing target-approaching and sensing-neighbor collision-free subtasks, and incorporating these subtasks into the constraints in an online constraint-based optimization framework, the proposed LTTE can systematically guarantee long-term target fencing under changing environments in the n-dimensional Euclidean space. Rigorous analysis is provided to guarantee asymptotical convergence with challenging nonlinear couplings induced by time-varying collision-free constraints. Finally, 3D simulations with changing environmental elements are conducted to demonstrate the effectiveness of the proposed algorithm.

In Chap. 6, we propose an equal-distance surrounding controller for robots to fence multiple moving targets with guaranteed collision avoidance. First, a distributed estimator is developed to approximate the center of moving targets. Then, an adaptive distributed control law is designed for the robots to accomplish equal-distance surrounding collaboratively. In particular, conditions for assuring asymptotical stability for the closed-loop system are derived. Finally, experimental results with unmanned surface vessels are reported to substantiate the effectiveness of the proposed algorithm.

In Chap. 7, we propose the flexible subgrouping and ordering coordination (FSOC) algorithm that enables a team of robots to cooperatively fence multiple scattered targets (MSTs) of different moving velocities, where each target is fenced into the interior of distinct convex hulls formed by different subgroups of robots. In particular, we first encode the three subtasks of flexible subgrouping, target convergence, and neighboring collision avoidance into control barrier function (CBF) constraints, and then formulate the FSOC problem as a decentralized constraint-based task-execution optimization one. Rigorous analysis is provided to guarantee the asymptotic convergence. Finally, 2D and 3D simulations have been conducted to confirm the effectiveness, robustness, adaptability, and scalability of the algorithm.

In Chap. 8, we propose a distributed bearing-only controller for multiple unmanned surface vessels (USVs) to fence and rotate evenly around a motional target with inter-USV topologies. The distributed control law consists of three terms, i.e., a bearing-only estimation term to approximate the target state, an upper-level surrounding term to fulfill the target-fencing mission, and a single-vessel regulation term to track the upper-level signal. Significantly, technical conditions are derived to guarantee the asymptotical stability of the closed-loop system. Finally, experimental results on a real platform and a target vessel are conducted to substantiate the effectiveness of the algorithm.

In Chap. 9, we summarize the whole monograph and further discuss future research topics of coordinated fencing missions.

References

1. T. Arai, E. Pagello, L.E. Parker, others: advances in multi-robot systems. IEEE Trans. Robot. Autom. **18**(5), 655–661 (2002)
2. G. Beni, J. Wang, Swarm intelligence in cellular robotic systems, in *Robots and Biological Systems: Towards a New Bionics?*. (Springer, 1993), pp. 703–712
3. E. Bonabeau, M. Dorigo, G. Theraulaz, *Swarm Intelligence: From Natural to Artificial Systems* (Oxford University Press, 1999)
4. C.W. Reynolds, flocks, herds, and schools: a distributed behavioral model. Comput. Graph. **21**(4) (1987)
5. T. Vicsek, A. Czirók, E. Ben-Jacob, I. Cohen, O. Shochet, Novel type of phase transition in a system of self-driven particles. Phys. Rev. Lett. **75**(6), 1226 (1995)
6. Z. Cheng, Z. Chen, T. Vicsek, D. Chen, H. Zhang, Pattern phase transitions of self-propelled particles: gases, crystals, liquids, and mills. New J. Phys. **18**(10), 103005 (2016)
7. I.D. Couzin, J. Krause, R. James, G.D. Ruxton, N.R. Franks, Collective memory and spatial sorting in animal groups. J. Theor. Biol. **218**(1), 1–11 (2002)
8. A. Cavagna, A. Cimarelli, I. Giardina, G. Parisi, R. Santagati, F. Stefanini, M. Viale, Scale-free correlations in starling flocks. Proc. Natl. Acad. Sci. **107**(26), 11865–11870 (2010)
9. M. Rubenstein, A. Cornejo, R. Nagpal, Programmable self-assembly in a thousand-robot swarm. Science **345**(6198), 795–799 (2014)
10. G. Vásárhelyi, C. Virágh, G. Somorjai, T. Nepusz, A.E. Eiben, T. Vicsek, Optimized flocking of autonomous drones in confined environments. Sci. Robot. **3**(20), eaat3536 (2018)
11. X. Zhou, X. Wen, Z. Wang, Y. Gao, H. Li, Q. Wang, T. Yang, H. Lu, Y. Cao, C. Xu et al., Swarm of micro flying robots in the wild. Sci. Robot. **7**(66), eabm5954 (2022)
12. J. Li, Development and application of surface unmanned boats. Firepower Comm. Control **37**(6), 203–207 (2012)
13. H. Tan, S. Cai, The next wave of the Singapore Navy: USVs. PIONEER (2021). https://www.mindef.gov.sg/web/portal/pioneer/article/regular-article-detail/ops-and-training/2021-Q1/chinese/10mar21_news-zh
14. L. Xu, Turkiye's "largest-ever" maritime exercise ends, with drones and unmanned boats taking the spotlight. China News Inf. Netwo. (2021). http://www.cnnewsinfo.com/202106/38o561.html

15. B.-B. Hu, H.-T. Zhang, B. Liu, H. Meng, G. Chen, Distributed surrounding control of multiple unmanned surface vessels with varying interconnection topologies. IEEE Trans. Control Syst. Technol. **30**(1), 400–407 (2021)
16. H. Zhu, *Polar Walker Ocean Exploration USV* (Zhuhai Yunzhou Company, 2021). https://www.afzhan.com/st211999/product_12486641.html
17. P. Michelucci, J.L. Dickinson, The power of crowds. Science **351**(6268), 32–33 (2016)
18. R. Olfati-Saber, R.M. Murray, Consensus problems in networks of agents with switching topology and time-delays. IEEE Trans. Autom. Control **49**(9), 1520–1533 (2004)
19. S. Wang, J. Duan, D. Shi, C. Xu, H. Li, R. Diao, Z. Wang, A data-driven multi-agent autonomous voltage control framework using deep reinforcement learning. IEEE Trans. Power Syst. **35**(6), 4644–4654 (2020)
20. D. Li, S.S. Ge, W. He, G. Ma, L. Xie, Multilayer formation control of multi-agent systems. Autom. **109**, 108558 (2019)
21. Z. Wang, M. Gombolay, Learning scheduling policies for multi-robot coordination with graph attention networks. IEEE Robot. Autom. Lett. **5**(3), 4509–4516 (2020)
22. H. Wei, V.R. Aparow, B.-B. Hu, C. Lv, A hierarchical distributed coordination framework for flexible and resilient vehicle platooning. IEEE Trans. Intell. Transp. Syst. **25**(6), 6090–6105 (2024)
23. H. Wei, B.-B. Hu, Y. Wang, C. Lv, Scalable and constrained consensus in multiagent systems: distributed model predictive control-based approaches. IEEE Trans. Ind. Inf. **20**(4), 5969–5978 (2024)
24. A. Farinelli, E. Zanotto, E. Pagello, and other, Advanced approaches for multi-robot coordination in logistic scenarios. Robot. Auton. Syst. **90**, 34–44 (2017)
25. B.-B. Hu, H.-T. Zhang, B. Liu, J. Ding, Y. Xu, C. Luo, H. Cao, Coordinated navigation control of cross-domain unmanned systems via guiding vector fields. IEEE Trans. Control Syst. Technol. **32**(2), 550–563 (2024)
26. A. Latha, S. Prasanna, S. Hemalatha, B. Sivakumar, A harmonized trust assisted energy efficient data aggregation scheme for distributed sensor networks. Cogn. Syst. Res. **56**, 14–22 (2019)
27. H. Yu, K. Song, J. Yang, C. Wu, K. Zhong, W. Lv, Modeling and research on power system of distributed sensor networks for long streamers in Marine seismic exploration. Sensors **20**(1), 28 (2020)
28. X. Ge, Q.-L. Han, M. Zhong, X.-M. Zhang, Distributed Krein space-based attack detection over sensor networks under deception attacks. Automatica **109**, 108557 (2019)
29. M. Hassan, D. Liu, D. Xu, A two-stage approach to collaborative fiber placement through coordination of multiple autonomous industrial robots. J. Intell. Robot. Syst. **95**(3), 915–933 (2019)
30. B.-B. Hu, H.-T. Zhang, W. Yao, J. Ding, M. Cao, Spontaneous-ordering platoon control for multirobot path navigation using guiding vector fields. IEEE Trans. Robot. **39**(4), 2654–2668 (2023)
31. B.-B. Hu, H.-T. Zhang, W. Yao, Z. Sun, M. Cao, Coordinated guiding vector field design for ordering-flexible multi-robot surface navigation. IEEE Trans. Autom. Control in press (2024). https://doi.org/10.1109/TAC.2024.3355321
32. M. Yan, X. Ai, M. Shahidehpour, Z. Li, J. Wen, S. Bahramira, A. Paaso, Enhancing the transmission grid resilience in ice storms by optimal coordination of power system schedule with pre-positioning and routing of mobile DC de-icing devices. IEEE Trans. Power Syst. **34**(4), 2663–2674 (2019)
33. H. Silva-Saravia, H. Pulgar-Painemal, D.A. Schoenwald, W. Ju, Adaptive coordination of damping controllers for enhanced power system stability. IEEE Open Access J. Power Energy **7**, 265–275 (2020)
34. C. Zhao, J. He, P. Cheng, J. Chen, Consensus-based energy management in smart grid with transmission losses and directed communication. IEEE Trans. Smart Grid **8**(5), 2049–2061 (2016)
35. N. Xing, H.-T. Zhang, L. Zhu, Prescribed-time collective evader-capturing for autonomous surface vehicles. Automatica **167**, 111761 (2024)

36. J. Xi, L. Wang, J. Zheng, X. Yang, Energy-constraint formation for multiagent systems with switching interaction topologies. IEEE Trans. Circuits Syst. I Regul. Pap. **67**(7), 2442–2454 (2020)
37. S.-L. Dai, S. He, X. Chen, X. Jin, Adaptive leader-follower formation control of nonholonomic mobile robots with prescribed transient and steady-state performance. IEEE Trans. Ind. Inform. **16**(6), 3662–3671 (2019)
38. J. Lin, Z. Miao, H. Zhong, W. Peng, Y. Wang, R. Fierro, Adaptive image-based leader-follower formation control of mobile robots with visibility constraints. IEEE Trans. Ind. Electron. **68**(7), 6010–6019 (2020)
39. Z. Li, H. Tnunay, S. Zhao, W. Meng, S.Q. Xie, Z. Ding, Bearing-only formation control with prespecified convergence time. IEEE Trans. Cybern. **52**(1), 620–629 (2021)
40. E. Nuno, T. Henández, M. Maghenem, A. Loría, E. Panteley, Leaderless consensus-based formation control of multiple nonholonomic mobile robots with interconnecting delays, in *Proceedings of the American Control Conference* (IEEE, 2019), pp. 4659–4664
41. Z. Lin, W. Ding, G. Yan, C. Yu, A. Giua, Leader-follower formation via complex Laplacian. Automatica **49**(6), 1900–1906 (2013)
42. M.H. Trinh, Q. Van Tran, D. Van Vu, P.D. Nguyen, H.-S. Ahn, Robust tracking control of bearing-constrained leader-follower formation. Automatica **131**, 109733 (2021)
43. D. Wang, W. Wang, Necessary and sufficient conditions for containment control of multi-agent systems with time delay. Automatica **103**, 418–423 (2019)
44. F. Chen, W. Ren, Y. Cao, Surrounding control in cooperative agent networks. Syst. Control Lett. **59**(11), 704–712 (2010)
45. Y. Cao, W. Ren, M. Egerstedt, Distributed containment control with multiple stationary or dynamic leaders in fixed and switching directed networks. Automatica **48**(8), 1586–1597 (2012)
46. Q. Xiong, P. Lin, Z. Chen, W. Ren, W. Gui, Distributed containment control for first-order and second-order multiagent systems with arbitrarily bounded delays. Int. J. Robust Nonlinear Control **29**(4), 1122–1131 (2019)
47. J. Qin, W.X. Zheng, H. Gao, Q. Ma, W. Fu, Containment control for second-order multiagent systems communicating over heterogeneous networks. IEEE Trans. Neural Netw. Learn. Syst. **28**(9), 2143–2155 (2016)
48. N. Gu, D. Wang, Z. Peng, L. Liu, Distributed containment maneuvering of uncertain underactuated unmanned surface vehicles guided by multiple virtual leaders with a formation. Ocean Eng. **187**, 105996 (2019)
49. J.A. Marshall, M.E. Broucke, B.A. Francis, Formations of vehicles in cyclic pursuit. IEEE Trans. Autom. Control **49**(11), 1963–1974 (2004)
50. T.-H. Kim, T. Sugie, Cooperative control for target-capturing task based on a cyclic pursuit strategy. Automatica **43**(8), 1426–1431 (2007)
51. J. Guo, G. Yan, Z. Lin, Local control strategy for moving-target enclosing under dynamically changing network topology. Syst. Control Lett. **59**(10), 654–661 (2010)
52. R. Zheng, Y. Liu, D. Sun, Enclosing a target by nonholonomic mobile robots with bearing-only measurements. Automatica **53**, 400–407 (2015)
53. R. Sharma, M. Kothari, C.N. Taylor, I. Postlethwaite, Cooperative target-capturing with inaccurate target information, in *Proceedings of the American Control Conference*, pp. 5520–5525 (2010)
54. X. Yu, L. Liu, Distributed circular formation control of ring-networked nonholonomic vehicles. Automatica **68**, 92–99 (2016)
55. X. Yu, L. Liu, Cooperative control for moving-target circular formation of nonholonomic vehicles. IEEE Trans. Autom. Control **62**(7), 3448–3454 (2016)
56. X. Yu, N. Ding, A. Zhang, H. Qian, Cooperative moving-target enclosing of networked vehicles with constant linear velocities. IEEE Trans. Cybern. **50**(2), 798–809 (2018)
57. L. Shi, R. Zheng, M. Liu, S. Zhang, Distributed control of cooperative multi-target enclosing by ring-coupled unicycles, in *Proceedings of the American Control Conference*, pp. 3523–3528 (2020)

58. L. Dou, C. Song, X. Wang, L. Liu, G. Feng, Target localization and enclosing control for networked mobile agents with bearing measurements. Automatica **118**, 109022 (2020)
59. L. Dou, X. Yu, L. Liu, X. Wang, G. Feng, Moving-target enclosing control for mobile agents with collision avoidance. IEEE Trans. Control Netw. Syst. **8**(4), 1669–1679 (2021)
60. Z. Yang, S. Zhu, C. Chen, G. Feng, X. Guan, Entrapping a target in an arbitrarily shaped orbit by a single robot using bearing measurements. Automatica **113**, 108805 (2020)
61. X. Dong, B. Yu, Z. Shi, Y. Zhong, Time-varying formation control for unmanned aerial vehicles: theories and applications. IEEE Trans. Control Syst. Technol. **23**(1), 340–348 (2014)
62. X. Dong, Y. Zhou, Z. Ren, Y. Zhong, Time-varying formation control for unmanned aerial vehicles with switching interaction topologies. Control Eng. Pract. **46**, 26–36 (2016)
63. Y. Jiang, Z. Peng, D. Wang, C.L.P. Chen, Line-of-sight target enclosing of an underactuated autonomous surface vehicle with experiment results. IEEE Trans. Ind. Inform. **16**(2), 832–841 (2019)
64. Z. Peng, Y. Jiang, J. Wang, Event-triggered dynamic surface control of an underactuated autonomous surface vehicle for target enclosing. IEEE Trans. Ind. Electron. **68**(4), 3402–3412 (2020)
65. B. Liu, Z. Chen, H.T. Zhang, X. Wang, T. Geng, H. Su, J. Zhao, Collective dynamics and control for multiple unmanned surface vessels. IEEE Trans. Control Syst. Technol. **28**(6), 2540–2547 (2019)
66. Z. Chen, A cooperative target-fencing protocol of multiple vehicles. Automatica **107**, 591–594 (2019)
67. L. Kou, Z. Chen, J. Xiang, Cooperative fencing control of multiple vehicles for a moving target with an unknown velocity. IEEE Trans. Autom. Control **67**(2), 1008–1015 (2021)
68. L. Kou, Y. Huang, Z. Chen, S. He, J. Xiang, Cooperative fencing control of multiple second-order vehicles for a moving target with and without velocity measurements. Int. J. Robust Nonlinear Control **31**(10), 4602–4615 (2021)
69. C. Wang, G. Xie, Limit-cycle-based decoupled design of circle formation control with collision avoidance for anonymous agents in a plane. IEEE Trans. Autom. Control **62**(12), 6560–6567 (2017)
70. B. Xu, H.T. Zhang, H. Meng, B.-B. Hu, D. Chen, G. Chen, Moving target surrounding control of linear multiagent systems with input saturation. IEEE Trans. Syst. Man Cybern. Syst. **52**(3), 1705–1715 (2020)
71. Y.J. Shi, R. Li, K.L. Teo, Cooperative enclosing control for multiple moving targets by a group of agents. Int. J. Control **88**(1), 80–89 (2015)
72. S. Shoja, M. Baradarannia, F. Hashemzadeh, M. Badamchizadeh, P. Bagheri, Surrounding control of nonlinear multi-agent systems with non-identical agents. ISA Trans. **70**, 219–227 (2017)
73. R. Li, Y. Shi, Y. Song, Multi-group coordination control of multi-agent system based on smoothing estimator. IET Control Theory & Appl. **10**(11), 1224–1230 (2016)
74. C. Li, L. Chen, Y. Guo, Y. Lyu, Cooperative surrounding control with collision avoidance for networked Lagrangian systems. J. Frank. Inst. **355**(12), 5182–5202 (2018)
75. K.H. Khalil, *Nonlinear Systems* (Prentice Hall, Upper Saddle River, 2002)

Part I
Single-Target Scenario

Chapter 2
Distributed Fencing Control via Neighboring Angle Repulsion

2.1 Introduction

In this chapter, we first consider the fundamental fencing problem for a constant-speed target under time-varying topologies. Although some previous works have addressed the fencing control of a constant-speed target, they are limited to solutions based on fixed topology or switching connected topologies, which are only effective in open and unobstructed environments. However, in increasingly complex and dynamically changing environments, it is important to further delve into the collaborative fencing theory and strategies for general time-varying topology, as well as the formation evolution during the fencing process. Therefore, this chapter proposes a distributed cooperative fencing controller via neighboring angle repulsion, which aims at solving the fencing problem of a constant-speed target under time-varying topologies İn what follows, we will introduce the related background.

The coordinated enclosing problem involves guiding a group of robots to enclose a target within their trajectories, offering various potential applications such as fleet convey, orbit maintenance, perimeter surveillance, and collaborative escort, which has seen significant research advancements in recent years. For example, a distributed method for enclosing a moving target using relative neighboring positions was developed in [1]. Similarly, a distributed cyclic controller was introduced in [2] to form specific pursuit patterns and encircle a moving target in 3D space. Subsequently, a local bearing-only protocol was proposed in [3] to achieve target encirclement. Notably, these studies assume that all robots have access to the target's information. Moreover, some efforts have been devoted to more realistic scenarios where only partial information about the target is available. Reference [4] proposed a robust protocol for encircling a target with limited information. Additional methods for enclosing a target with partial knowledge can be found in [5, 6], focusing on fixed and switching topologies, respectively. It is worth mentioning that existing solutions for the target-enclosing problem are applicable to both stationary and moving targets.

B.-B. Hu and H.-T. Zhang, *Multi-Robot Systems*, Cognitive Intelligence and Robotics,
https://doi.org/10.1007/978-981-97-9214-6_2

Different from the previous target-enclosing methods focusing on the trajectories, another approach named the target-fencing problem can govern the robots to collectively fence a target within their convex hull at all times. A common technique for addressing the target-fencing control problem relies on strategies designed for labeled robots, where the stand-off displacement for each robot is predefined. As the foundational work, an estimation and control framework was introduced in [7] to enable multiple robots to fence stationary targets. Later, a fencing controller was proposed in [8] for linear second-order MASs. For more complex nonlinear and non-identical MASs, an adaptive controller was developed in [9]. Some other efforts based on labeled robots were explored in [10–12]. However, these strategies for labeled robots have inherent limitations, prompting the need for approaches that do not rely on labeled robots. Consequently, a limit-cycle-based strategy was devised in [13] to achieve a circular formation around a target. Furthermore, a hierarchical control structure was proposed in [14] to fence a specified target. Later, a cooperative controller incorporating attractive, repulsive, and rotational components was designed in [15] to fence a target without requiring a predefined distance.

However, the aforementioned labeled algorithms [7–12] always predefine the relative distribution and steady states of different robots, which limits the algorithm's effectiveness in handling more complex cooperative tasks. Despite some approaches in [13–15] introducing the usage of unlabeled robots, they are tailored for first-order multi-agent systems (MASs) with a static target. Additionally, due to the challenges in the distributed calculation of steady states, the proposed algorithms in [13–15] do not guarantee a specific formation pattern. To tackle these problems, we proposed a distributed fencing controller via neighboring angle repulsion, which not only addresses the higher order fencing problem but also achieves a regular polygon formation with collision avoidance. Overall, the main contribution of this chapter is three-fold.

1. Develop a distributed fencing controller to enable a group of unlabeled robots to simultaneously achieve moving target fencing, maintain a regular polygon formation, and ensure collision avoidance.
2. Address time-varying network topologies through the use of nearest angle rules, achieve collision avoidance through distance regulation and angle repulsion, and mitigate the effects of target velocity estimation errors on the closed-loop system.
3. Rigorously prove the convergence of a complex nonlinear networked system.

The rest of this chapter is given below. Section 2.2 introduces the problem formulation and presents the explicit controller and the main objectives. In Sect. 2.3, the main theorem is rigorously proved. Section 2.4 provides numerical simulations to demonstrate the effectiveness. Finally, we draw the conclusion in Sect. 2.5.

2.2 Problem Formulation of Fencing a Constant-Velocity Target

In this section, we firstly introduce a team of n robots represented by $\mathcal{V} = \{1, 2, \ldots, n\}$, each of which is governed by second-order dynamics as below,

$$\begin{aligned} \dot{x}_i(t) &= v_i(t) \\ \dot{v}_i(t) &= u_i(t),\ \ i \in \mathcal{V}, \end{aligned} \tag{2.1}$$

where $x_i(t), v_i(t) \in \mathbb{R}^2$ denote the position and velocity of robot i, respectively, and $u_i(t) \in \mathbb{R}^2$ is the control input.

Then, we consider a moving target $x_d(t), v_d(t) \in \mathbb{R}^2$ as

$$\begin{aligned} \dot{x}_d(t) &= v_d(t) \\ \dot{v}_d(t) &= u_d(t). \end{aligned} \tag{2.2}$$

We assume that the position $x_d(t)$ is measurable and available to the followers, but the velocity $v_d(t)$ is not. Let $\widehat{v}_i(t)$ and $\widehat{u}_i(t) = \dot{\widehat{v}}_i(t)$ be the estimation of $v_d(t)$ and $u_d(t)$ by robot i, respectively, which is usually generated by a dynamic observer

$$\begin{aligned} [\widehat{v}_i(t), \widehat{u}_i(t)] &= F(\zeta_i(t), x_d(t)) \\ \dot{\zeta}_i(t) &= G(\zeta_i(t), x_d(t)),\ \ i \in \mathcal{V}. \end{aligned} \tag{2.3}$$

Here, F, G are some functions to be designed later. Note that the observer (2.3) aims to drive the estimation error to zero in the sense that

$$\lim_{t\to\infty} \widehat{v}_i(t) - v_d(t) = 0,\ \ \lim_{t\to\infty} \widehat{u}_i(t) - u_d(t) = 0,\ \ i \in \mathcal{V} \tag{2.4}$$

exponentially. We assume that $v_d(t)$ and $u_d(t)$ are bounded, also $u_d(t)$ and $\widehat{u}_i(t)$ are uniformly continuous in t, which is also reasonable in practical scenarios.

We define e_i as the relative position between each robot $i \in \mathcal{V}$ and the target as

$$e_i(t) = x_i(t) - x_d(t), \tag{2.5}$$

which can be rewritten in the polar coordinates

$$e_i(t) = \rho_i(t) s_{\theta_i(t)}. \tag{2.6}$$

Meanwhile, let $\widehat{\delta}_i(t) = v_i(t) - \widehat{v}_i(t)$ as

$$\begin{bmatrix} \widehat{\eta}_i(t) \\ \rho_i(t)\widehat{\omega}_i(t) \end{bmatrix} = J_{\theta_i(t)}^{\mathsf{T}} \widehat{\delta}_i(t), \tag{2.7}$$

which can also be reformulated in the polar coordinates with $\widehat{\eta}_i(t)$ and $\widehat{\omega}_i(t)$. To avoid confusion, we leverage the following notations for convenience,

$$s_{\theta_i} = \begin{bmatrix} \cos\theta_i \\ \sin\theta_i \end{bmatrix}, \ t_{\theta_i} = \begin{bmatrix} -\sin\theta_i \\ \cos\theta_i \end{bmatrix},$$

and

$$J_{\theta_i} = \begin{bmatrix} s_{\theta_i} & t_{\theta_i} \end{bmatrix}, \ J'_{\theta_i} = \frac{\partial J_{\theta_i}}{\theta_i} = \begin{bmatrix} t_{\theta_i} & -s_{\theta_i} \end{bmatrix}. \tag{2.8}$$

Moreover, let

$$\theta_{i,k} = \theta_i - \theta_k + 2\kappa\pi \in (-\pi, \pi] \tag{2.9}$$

for an appropriately selected integer κ. According to the above introduction, the position states x_i, x_d are available to each robot i, and hence ρ_i, θ_i, and the velocity states $v_i, \widehat{v}_i$, which implies that $\widehat{\eta}_i(t), \widehat{\omega}_i(t)$ are available to each robot as well.

Therefore, a distributed fencing controller for each robot $i, i \in \mathcal{V}$, can be designed as

$$u_i = \widehat{u}_i + J'_{\theta_i} \begin{bmatrix} \widehat{\eta}_i\widehat{\omega}_i \\ \rho_i\widehat{\omega}_i^2 \end{bmatrix} + J_{\theta_i} \begin{bmatrix} -k_1\alpha(\rho_i) - k_2\widehat{\eta}_i \\ \widehat{\eta}_i\widehat{\omega}_i + \rho_i(-k_3 \sum\limits_{k\in\mathcal{N}_i} \beta(\theta_{i,k}) - k_4\widehat{\omega}_i) \end{bmatrix}. \tag{2.10}$$

Here, the argument (t) is omitted for conveniences, $k_1, k_2, k_3, k_4 > 0$ represents the four positive parameters, $\gamma > 0$ denotes a sensing radius, $\rho_o > 0$ is a predefined desired distance between each robot and the target, and $0 < M < \rho_o$ denotes a specified minimum (safe) distance between the robots and the target. Moreover, it follows from (2.10) that the aforementioned two functions α and β are defined as

$$\alpha(\rho_i) = \frac{(\rho_i - \rho_o)(\rho_o - M)}{(\rho_i - M)^3}, \beta(\theta_{i,k}) = \frac{2\gamma(|\theta_{i,k}| - \gamma)\theta_{i,k}}{|\theta_{i,k}|^4},$$

where the sensing neighborhood is

$$\mathcal{N}_i(t) = \{k \in \mathcal{V}, k \neq i \mid |\theta_{i,k}(t)| < \gamma\}. \tag{2.11}$$

Now, the main problem addressed in this paper is to achieve the three following objectives for closed-loop system composed of (2.1), (2.3), and (2.10): (1) **P1**: $\lim_{t\to\infty} \rho_i(t) = \rho_o, \forall i \in \mathcal{V}$, which implies that the robot–target distance asymptotically converges to ρ_o. (2) **P2**: $\lim_{t\to\infty} |\theta_{i,k}(t)| \geq 2\pi/n, \ \forall i \neq k \in \mathcal{V}$, which implies that all the robots asymptotically maintain an equal angle distribution. (3) **P3**: $\rho_i(t) > M, \ \theta_{i,k}(t) \neq 0, \ \forall t \geq 0, \ \forall i \neq k \in \mathcal{V}$, which implies that collisions among the robots and the target are all guaranteed.

Remark 2.1 Given properties P1 and P2, a finite time T exists in which the target-fencing property $P_{x_d(t)}(x(t)) = 0,\ \forall t \geq T$ can be achieved, where $x = [x_1, \ldots, x_n]^\top$. In other words, the target x_d remains inside the convex hull formed by the robots, with $P_{x_d}(x) = \min_{s \in \mathrm{co}(x)} |x_d - s|$:

$$\mathrm{co}(x) = \left\{ \sum_{i \in \mathcal{V}} \lambda_i x_i : \lambda_i \geq 0, \forall i \text{ and } \sum_{i \in \mathcal{V}} \lambda_i = 1 \right\}. \tag{2.12}$$

Various target-fencing strategies have been explored in the literature, such as in [7–15]. A notable aspect of properties P1 and P2 is that the robots gradually converge to a regular polygon formation around the target, centered at a radius ρ_o, with an equal angle of $2\pi/n$ between each pair of adjacent robots. Property P3 ensures collision avoidance not only between the robots themselves but also between each robot and the target.

To achieve P1–P3, we also require the following three conditions. (1) **C1**: The observer in (2.3) is designed so that the estimation error in (2.4) converges exponentially. (2) **C2**: The initial positions of the robots and the target meet the conditions $\rho_i(0) > M$ and $\theta_{i,k}(0) \neq 0$ for all $i \neq k \in \mathcal{V}$. (3) **C3**: The sensing range parameter is constrained by $2\pi/n \leq \gamma \leq 4\pi/n$.

Remark 2.2 Condition C1 refers to the existence of a local estimator for the moving target's velocity based on position measurements by each robot. Each robot designed this estimator locally without requiring network communication. Such kind of problem has been extensively studied in the literature, for instance, using output regulation theory [16, 17] if u_d follows a specific dynamics. Moreover, to provide a complete design, an example is included for the case where the target velocity is constant with $u_d = 0$. Let $\zeta_i = [\widehat{x}_i, \widehat{v}_i]^\top$, the observer in (2.3) can be designed to be

$$\begin{aligned} \dot{\widehat{x}}_i &= -\phi_1(\widehat{x}_i - x_d) + \widehat{v}_i \\ \dot{\widehat{v}}_i &= \widehat{u}_i = -\phi_2\phi_1(\widehat{x}_i - x_d),\ i \in \mathcal{V} \end{aligned} \tag{2.13}$$

with $\phi_1, \phi_2 > 0$. Let $\widetilde{\zeta}_i = \zeta_i - [x_d, v_d]^\top$. One has $\dot{\widetilde{\zeta}}_i = A\widetilde{\zeta}_i$ for a Hurwitz matrix

$$A = \begin{bmatrix} -\phi_1 & 1 \\ -\phi_2\phi_1 & 0 \end{bmatrix},$$

which implies $\lim_{t\to\infty} \widetilde{\zeta}_i(t) = 0$ and the condition in (2.4) is satisfied.

Remark 2.3 Condition C2 is essential for ensuring the collision avoidance property P3, which requires robots to begin from a non-colliding, generic distribution. Condition C3 requires that $\gamma \geq 2\pi/n$ to ensure the equal angle distribution in P2. If $\gamma < 2\pi/n$, a trivial steady-state distribution can be found for $2\pi/n > \theta_{i,i+1} > \gamma$ for $i = 1, \ldots, n-1$, but $\theta_{n,1} > 2\pi/n > \gamma$, resulting in $\mathcal{N}_i = \emptyset$ for $i \in \mathcal{V}$ and violating P2. Additionally, the condition $\gamma \leq 4\pi/n$ in C3 provides an upper bound on the sensing angle range, simplifying the analysis of angle distribution. Such kind of constraint essentially ensures that no robot has more than two neighbors when the regular polygon formation is achieved.

The primary goal of this chapter is to demonstrate that the proposed controller (2.10) can achieve the three specified properties P1–P3 under the given conditions C1–C3. It is still worth mentioning two additional technical aspects as follows:

(i) **Controller Functionality**: The controller performs dual functions—maintaining the distance between robots and the target and regulating the angles between the robots. Distance maintenance ensures the cohesion of the robot group within a specified radius, while angle rules ensure the correct angular relationships among the robots. Importantly, the network topology is time-varying; a fixed, connected topology is not assumed, and robots are not explicitly labeled as seen in, e.g., [13–15].

(ii) **Target Velocity Estimation**: Each robot estimates the target's velocity, and while the estimation error converges to zero, it affects the system's behavior, particularly in systems with strong nonlinearities. Even in stable nonlinear systems, small disturbances can lead to divergence or finite escape times [18]. Therefore, assuming zero estimation error is not feasible, which will pose significant challenges in proving the main theorem.

2.3 Theoretical Convergence Analysis

In this section, we will present the main theorem with a rigorous proof.

Theorem 2.1 *The closed-loop system, consisting of* (2.1), (2.3), *and* (2.10), *successfully achieves properties P1, P2, and P3, provided that conditions C1, C2, and C3 are satisfied.*

Proof For clarity, we divide the proof into four steps. First, we perform some preliminary manipulation of the closed-loop system to select appropriate Lyapunov function candidates. In the subsequent steps, we provide proofs for properties P3, P1, and P2 in sequence.

(i) In Step 1, we will select a suitable Lyapunov function candidate. Let $\delta_i = \dot{e}_i$ be the velocity difference between robot i and the target, which follows from the definition of e_i in the Cartesian coordinates (2.5) that $\delta_i = v_i - v_d$. Meanwhile, in the polar coordinates (2.6), it can be deduced that

$$\delta_i = \rho_i t_{\theta_i} \dot{\theta}_i + \dot{\rho}_i s_{\theta_i} = J_{\theta_i} \begin{bmatrix} \eta_i \\ \rho_i \omega_i \end{bmatrix} \tag{2.14}$$

for $\eta_i = \dot{\rho}_i$ and $\omega_i = \dot{\theta}_i$. Accordingly, one has that the difference between $\widehat{\delta}_i$ in (2.7) and δ_i in (2.14) can be expressed in both coordinates as

$$\widehat{\delta}_i - \delta_i = v_d - \widehat{v}_i = J_{\theta_i} \begin{bmatrix} \widetilde{\eta}_i \\ \rho_i \widetilde{\omega}_i \end{bmatrix} \tag{2.15}$$

where $\widetilde{\eta}_i = \widehat{\eta}_i - \eta_i$ and $\widetilde{\omega}_i = \widehat{\omega}_i - \omega_i$. For the convenience of presentation, let $\widetilde{u}_i = \widehat{u}_i - u_d$. From the fact that J_{θ_i} in (2.8) is nonsingular, it, together with C1 and (2.15), gives that

$$\lim_{t\to\infty} \widetilde{\eta}_i(t) = 0, \ \lim_{t\to\infty} \rho_i(t)\widetilde{\omega}_i(t) = 0, \ \lim_{t\to\infty} \widetilde{u}_i(t) = 0. \tag{2.16}$$

Next, one has that $\dot{\delta}_i = \ddot{e}_i$. On the one hand,

$$\dot{\delta}_i = J'_{\theta_i} \begin{bmatrix} \eta_i \omega_i \\ \rho_i \omega_i^2 \end{bmatrix} + J_{\theta_i} \begin{bmatrix} \dot{\eta}_i \\ \eta_i \omega_i + \rho_i \dot{\omega}_i \end{bmatrix} \tag{2.17}$$

from (2.14). On the other hand, from the condition in (2.10), it can be deduced that

$$\begin{aligned} \dot{\delta}_i =& \dot{v}_i - \dot{v}_d = u_i - u_d \\ =& \widehat{u}_i - u_d + J'_{\theta_i} \begin{bmatrix} \widehat{\eta}_i \widehat{\omega}_i \\ \rho_i \widehat{\omega}_i^2 \end{bmatrix} + J_{\theta_i} \begin{bmatrix} -k_1\alpha(\rho_i) - k_2\widehat{\eta}_i \\ \widehat{\eta}_i \widehat{\omega}_i + \rho_i(-k_3 \sum_{k\in\mathcal{N}_i} \beta(\theta_{i,k}) - k_4\widehat{\omega}_i) \end{bmatrix}. \end{aligned} \tag{2.18}$$

Combining (2.17) and (2.18) together yields

$$\begin{bmatrix} \dot{\eta}_i \\ \rho_i \dot{\omega}_i \end{bmatrix} = \begin{bmatrix} -k_1\alpha(\rho_i) - k_2\widehat{\eta}_i \\ \widehat{\eta}_i \widehat{\omega}_i + \rho_i(-k_3 \sum_{k\in\mathcal{N}_i} \beta(\theta_{i,k}) - k_4\widehat{\omega}_i) - \eta_i\omega_i \end{bmatrix} + J_{\theta_i}^{\mathsf{T}} J'_{\theta_i} \begin{bmatrix} \widehat{\eta}_i \widehat{\omega}_i - \eta_i \omega_i \\ \rho_i \widehat{\omega}_i^2 - \rho_i \omega_i^2 \end{bmatrix} + J_{\theta_i}^{\mathsf{T}} \widetilde{u}_i.$$

Meanwhile, from the condition that

$$\begin{aligned} \widehat{\eta}_i \widehat{\omega}_i &= \eta_i \omega_i + \widetilde{\eta}_i \omega_i + \eta_i \widetilde{\omega}_i + \widetilde{\eta}_i \widetilde{\omega}_i \\ \widehat{\omega}_i^2 &= \omega_i^2 + 2\omega_i \widetilde{\omega}_i + \widetilde{\omega}_i^2, \end{aligned}$$

one has

$$\begin{bmatrix} \dot{\eta}_i \\ \dot{\omega}_i \end{bmatrix} = \begin{bmatrix} -k_1\alpha(\rho_i) - k_2\eta_i \\ -k_3 \sum_{k\in\mathcal{N}_i} \beta(\theta_{i,k}) - k_4\omega_i \end{bmatrix} + \xi_i$$

for

$$\begin{aligned} \xi_i = \begin{bmatrix} \epsilon_i \\ \varepsilon_i \end{bmatrix} =& \begin{bmatrix} -k_2\widetilde{\eta}_i \\ -k_4\widetilde{\omega}_i + (\widetilde{\eta}_i \omega_i + \eta_i \widetilde{\omega}_i + \widetilde{\eta}_i \widetilde{\omega}_i)/\rho_i \end{bmatrix} \\ &+ R_i^{-1} J_{\theta_i}^{\mathsf{T}} J'_{\theta_i} \begin{bmatrix} \widetilde{\eta}_i \omega_i + \eta_i \widetilde{\omega}_i + \widetilde{\eta}_i \widetilde{\omega}_i \\ 2\rho_i \omega_i \widetilde{\omega}_i + \rho_i \widetilde{\omega}_i^2 \end{bmatrix} + R_i^{-1} J_{\theta_i}^{\mathsf{T}} \widetilde{u}_i \end{aligned}$$

$$
\begin{aligned}
&= \begin{bmatrix} -k_2\widetilde{\eta}_i - 2\rho_i\omega_i\widetilde{\omega}_i - \rho_i\widetilde{\omega}_i^2 \\ -k_4\widetilde{\omega}_i + 2(\widetilde{\eta}_i\omega_i + \eta_i\widetilde{\omega}_i + \widetilde{\eta}_i\widetilde{\omega}_i)/\rho_i \end{bmatrix} + R_i^{-1} J_{\theta_i}^{\top}\widetilde{u}_i \\
&= \begin{bmatrix} -2\rho_i\omega_i\widetilde{\omega}_i \\ 2(\widetilde{\eta}_i\omega_i + \eta_i\widetilde{\omega}_i)/\rho_i \end{bmatrix} + \begin{bmatrix} \sigma_i \\ \varsigma_i \end{bmatrix},
\end{aligned} \tag{2.19}
$$

with $R_i = \text{diag}(1, \rho_i)$ and

$$
\begin{bmatrix} \sigma_i \\ \varsigma_i \end{bmatrix} = \begin{bmatrix} -k_2\widetilde{\eta}_i - \rho_i\widetilde{\omega}_i^2 \\ -k_4\widetilde{\omega}_i + 2(\widetilde{\eta}_i\widetilde{\omega}_i)/\rho_i \end{bmatrix} + R_i^{-1} J_{\theta_i}^{\top}\widetilde{u}_i. \tag{2.20}
$$

Accordingly, it can be deduced that the closed-loop system is composed of

$$
\begin{aligned}
\dot{\rho}_i &= \eta_i \\
\dot{\eta}_i &= -k_1\alpha(\rho_i) - k_2\eta_i + \epsilon_i,
\end{aligned} \tag{2.21}
$$

and

$$
\begin{aligned}
\dot{\theta}_i &= \omega_i \\
\dot{\omega}_i &= -k_3 \sum_{k\in\mathcal{N}_i} \beta(\theta_{i,k}) - k_4\omega_i + \varepsilon_i.
\end{aligned} \tag{2.22}
$$

Then, given the system (2.21), we first choose a function concerning ρ_i as

$$
P(\rho_i) = \left(\frac{\rho_i - \rho_o}{\rho_i - M}\right)^2, \quad \rho_i \in (M, \infty), \tag{2.23}
$$

which is continuously differentiable. We pick a Lyapunov function candidate

$$
V_1(\rho, \eta) = \frac{1}{2}\sum_{i=1}^{n} \left(k_1 P(\rho_i) + \eta_i^2\right) \tag{2.24}
$$

with $\rho = [\rho_1, \ldots, \rho_n]^{\top}$ and $\eta = [\eta_1, \ldots, \eta_n]^{\top}$. Taking the time derivative of $V_1(\rho, \eta)$ along the trajectories yields (2.21)

$$
\begin{aligned}
\dot{V}_1(\rho, \eta) &= \sum_{i=1}^{n} \left(\frac{k_1}{2}\frac{\partial P(\rho_i)}{\partial \rho_i}\dot{\rho}_i + \eta_i\dot{\eta}_i\right) \\
&= \sum_{i=1}^{n} \left(k_1\frac{(\rho_i - \rho_o)(\rho_o - M)}{(\rho_i - M)^3}\eta_i - \eta_i k_1\frac{(\rho_i - \rho_o)(\rho_o - M)}{(\rho_i - M)^3} - k_2\eta_i^2 + \epsilon_i\eta_i\right) \\
&= \sum_{i=1}^{n} \left(-k_2\eta_i^2 + \eta_i\epsilon_i\right) \\
&\le \sum_{i=1}^{n} \left(-\frac{k_2}{2}\eta_i^2 + \frac{1}{2k_2}\epsilon_i^2\right).
\end{aligned} \tag{2.25}
$$

Analogously, for the system (2.22), we also choose a function concerning $\theta_{i,k}$ as

$$U(\theta_{i,k}) = \begin{cases} 0, & |\theta_{i,k}| \in [\gamma, \pi] \\ \left(\frac{|\theta_{i,k}|-\gamma}{|\theta_{i,k}|}\right)^2, & |\theta_{i,k}| \in (0, \gamma) \end{cases}, \tag{2.26}$$

which is continuously differentiable. Taking the time derivative of (2.26) yields

$$\dot{U}(\theta_{i,k}) = \begin{cases} 0 & |\theta_{i,k}| \in [\gamma, \pi] \\ \beta(\theta_{i,k})(\dot{\theta}_i - \dot{\theta}_k), & |\theta_{i,k}| \in (0, \gamma) \end{cases}. \tag{2.27}$$

In this way, one has that $\dot{\theta}_{i,k} = \dot{\theta}_i - \dot{\theta}_k$ when $|\theta_{i,k}| \in (0, \gamma)$ with $\gamma < \pi$. Also, it can be deduced that $\beta(\theta_{i,k}) = -\beta(\theta_{k,i})$. Then, we can pick a Lyapunov function candidate as follows:

$$V_2(\theta, \omega) = \frac{k_3}{2} \sum_{i=1}^{n} \sum_{k=1}^{n} U(\theta_{i,k}) + \frac{1}{2} \sum_{i=1}^{n} \omega_i^2, \tag{2.28}$$

where $\theta = [\theta_1, \ldots, \theta_n]^{\mathrm{T}}$ and $\omega = [\omega_1, \ldots, \omega_n]^{\mathrm{T}}$. Taking the time derivative along the trajectories of (2.22) yields

$$\begin{aligned} \dot{V}_2(\theta, \omega) =& \frac{k_3}{2} \sum_{i=1}^{n} \sum_{k \in \mathcal{N}_i} \beta(\theta_{i,k})(\dot{\theta}_i - \dot{\theta}_k) + \sum_{i=1}^{n} \omega_i \dot{\omega}_i \\ =& \sum_{i=1}^{n} \omega_i \left(-k_4 \omega_i + \varepsilon_i\right) \\ \leq& \sum_{i=1}^{n} \left(-\frac{k_4}{2} \omega_i^2 + \frac{1}{2k_4} \varepsilon_i^2\right). \end{aligned} \tag{2.29}$$

It follows from (2.25) and (2.29) that

$$\begin{aligned} \dot{V}_1(\rho, \eta) + \dot{V}_2(\theta, \omega) \leq& \sum_{i=1}^{n} \left(-\frac{k_2}{2} \eta_i^2 - \frac{k_4}{2} \omega_i^2 + \frac{1}{2k_2} \epsilon_i^2 + \frac{1}{2k_4} \varepsilon_i^2\right) \\ \leq& \sum_{i=1}^{n} \left(-a_i \eta_i^2 - b_i \omega_i^2 + c_i\right), \end{aligned} \tag{2.30}$$

for

$$a_i = \frac{k_2}{2} - \frac{6\widetilde{\omega}_i^2}{k_4 \rho_i^2}, \; b_i = \frac{k_4}{2} - \frac{4(\rho_i \widetilde{\omega}_i)^2}{k_2} - \frac{6\widetilde{\eta}_i^2}{k_4 \rho_i^2}, \; c_i = \frac{\sigma_i^2}{k_2} + \frac{3\varsigma_i^2}{2k_4}$$

on the space

$$\mathbb{S} = \left\{ [\rho^\top, \eta^\top, \theta^\top, \omega^\top]^\top \in \mathbb{R}^{4n} \mid \rho_i \in (M, \infty), |\theta_{i,k}| \in (0, \pi], \ i \neq k \in \mathcal{V} \right\}.$$

It is noted that the following conditions:

$$\begin{aligned} \epsilon_i^2 =&(-2\rho_i\omega_i\widetilde{\omega}_i + \sigma_i)^2 \le 8(\rho_i\widetilde{\omega}_i)^2\omega_i^2 + 2\sigma_i^2 \\ \varepsilon_i^2 =&(2\widetilde{\eta}_i\omega_i/\rho_i + 2\eta_i\widetilde{\omega}_i/\rho_i + \varsigma_i)^2 \\ \le& 12(\widetilde{\eta}_i/\rho_i)^2\omega_i^2 + 12(\widetilde{\omega}_i/\rho_i)^2\eta_i^2 + 3\varsigma_i^2 \end{aligned}$$

are leveraged in the calculation.

(ii) In Step 2, we will prove the property of P3. First of all, the property P3 is equivalent to the claim that the states always stay on the space $\mathbb{S}$ for $t \geq 0$. Under C2, the states are initially on space $\mathbb{S}$ for $t = 0$. If P3 does not hold, there exists $\tau > 0$ such that the states are on the space $\mathbb{S}$ for $t \in [0, \tau)$ but not at $t = \tau$. A contradiction is then found below.

For $t \in [0, \tau)$, the states stay on $\mathbb{S}$ and $\rho_i \in (M, \infty)$, then, by (2.16), there exist two real numbers ν_1 (not necessarily positive) and $\nu_2 > 0$ such that $a_i, b_i > \nu_1$ and $\sum_{i=1}^n c_i \leq \nu_2$.

Let $V(t) = V_1(\rho(t), \eta(t)) + V_2(\theta(t), \omega(t))$. Then, it follows from (2.24), (2.28), and (2.30) that the time derivative of $V(t)$ satisfies

$$\dot{V} \leq \sum_{i=1}^{n} \left(-a_i\eta_i^2 - b_i\omega_i^2 + c_i\right) \leq -\nu_1 V + \nu_2.$$

According to the comparison principle [19], it can be deduced that

$$V(\tau) \leq e^{-\nu_1\tau}V(0) + \nu_2 \int_0^\tau e^{-\nu_1(\tau-s)}ds.$$

Since $V(0)$ is bounded under the condition C2, one has that $V(\tau)$ is bounded, which further implies that the states stay on $\mathbb{S}$ at $t = \tau$. The property P3 is thus proved.

(iii) In Step 3, we will prove the property P1. From the previous Step 2, one has that $\rho_i \in (M, \infty)$ for all $t \geq 0$, which together with (2.16) gives that

$$\lim_{t\to\infty} \widetilde{\omega}_i(t) = 0. \tag{2.31}$$

It follows from the definitions of σ_i, ς_i in (2.20), and the conditions in (2.16) and (2.31) that

$$\lim_{t\to\infty} \sigma_i(t) = 0, \ \lim_{t\to\infty} \varsigma_i(t) = 0. \tag{2.32}$$

Bearing in mind (2.16), (2.31), and (2.32), it can be deduced that $a_i > 0, b_i > 0$, and $\lim_{t\to\infty} c_i(t) = 0$ after a finite time T. Note that all the convergences a_i, b_i, c_i hold exponentially. Next, it can be deduced that

$$V(t) \leq V(T) - \sum_{i=1}^{n} \int_T^t \left(a_i(s)\eta_i^2(s) + b_i(s)\omega_i^2(s)\right) ds + \sum_{i=1}^{n} \int_T^t c_i(s)ds, \ \forall t \geq T.$$

Since $\lim_{t\to\infty} c_i(t) = 0$ exponentially, it can be deduced that $\sum_{i=1}^{n} \int_T^t c_i(s)ds$ is bounded, so is $V(t)$ for $t \geq T$. In this way, one has that $\eta(t)$ and $\omega(t)$ are bounded. Also, it, together with (2.16), (2.31), and (2.32), gives that $\epsilon_i, \varepsilon_i$ in (2.19) fulfill

$$\lim_{t\to\infty} \epsilon_i(t) = 0, \ \lim_{t\to\infty} \varepsilon_i(t) = 0. \tag{2.33}$$

According to the fact that $V(t)$ is bounded, it, together with (2.21), (2.22), (2.24), and (2.28), gives that $\eta_i, \omega_i, \dot{\eta}_i, \dot{\omega}_i$ are bounded. Moreover, given the condition that $\dot{v}_d(t) - \dot{\widehat{v}}_i(t) = u_d(t) - \widehat{u}_i(t)$ is uniformly continuous in t, it follows from (2.4) and (2.15) that

$$\lim_{t\to\infty} \left[\dot{\widehat{\delta}}_i(t) - \dot{\delta}_i(t)\right] = \lim_{t\to\infty} \left[\dot{v}_d(t) - \dot{\widehat{v}}_i(t)\right] = 0 \tag{2.34}$$

based on Barbalat's lemma. Given the condition of (2.15) again, it can be deduced that

$$\dot{\widehat{\delta}}_i - \dot{\delta}_i = J_{\theta_i}' \omega_i \begin{bmatrix} \widetilde{\eta}_i \\ \rho_i \widetilde{\omega}_i \end{bmatrix} + J_{\theta_i} \begin{bmatrix} \dot{\widetilde{\eta}}_i \\ \eta_i \widetilde{\omega}_i + \rho_i \dot{\widetilde{\omega}}_i \end{bmatrix}, \tag{2.35}$$

which, together with (2.34), implies that $\dot{\widetilde{\eta}}_i, \rho_i \dot{\widetilde{\omega}}_i$ are bounded, so is $\dot{\widetilde{\omega}}_i$. Then, one has that $\widehat{\eta}_i, \widehat{\omega}_i, \dot{\widehat{\eta}}_i, \dot{\widehat{\omega}}_i$ are bounded.

Let $d_i = a_i\eta_i^2 + b_i\omega_i^2$. From the fact that $\int_T^t d_i(s)ds$ is upper bounded and $\lim_{t\to\infty} \int_T^t d_i(s)ds$ exists, one has that $\dot{d}_i$ contains $a_i, b_i, \eta_i, \omega_i, \dot{\eta}_i, \dot{\omega}_i, \rho_i^{-1}, \widetilde{\omega}_i, \dot{\widetilde{\omega}}_i$, $\rho_i\widetilde{\omega}_i, \rho_i\dot{\widetilde{\omega}}_i, \widetilde{\eta}_i, \dot{\widetilde{\eta}}_i$ which are all bounded, so is $\dot{d}_i$. Then, it can be deduced that $d_i(t)$ is uniformly continuous in t. Using Barbalat's lemma, one has that $\lim_{t\to\infty} d_i(t) = 0$ and hence $\lim_{t\to\infty} \eta_i(t) = 0$ and $\lim_{t\to\infty} \omega_i(t) = 0$ when a_i and b_i are lower bound away from zero.

For the case of $\rho_i \in (M, \infty)$, one has that $\dot{\alpha}(\rho_i)$ is bounded. It follows from (2.19) and (2.21) that $\ddot{\eta}_i$ contains $\dot{\alpha}(\rho_i), \dot{\eta}_i, \eta_i, \omega_i, \dot{\omega}_i, \rho_i\widetilde{\omega}_i, \widetilde{\omega}_i, \rho_i\dot{\widetilde{\omega}}_i, \dot{\widetilde{\eta}}_i, \dot{\widetilde{u}}_i$ which are all bounded, so is $\ddot{\eta}_i$. It implies that $\dot{\eta}_i(t)$ is uniformly continuous in t. Also, it has been proved that $\lim_{t\to\infty} \eta_i(t) = 0$. Using Barbalat's lemma, one has $\lim_{t\to\infty} \dot{\eta}_i(t) = 0$. It, together with (2.21), gives $\lim_{t\to\infty} \alpha(\rho_i(t)) = 0$, which further implies $\lim_{t\to\infty} \rho_i(t) = \rho_o, \forall i \in \mathcal{V}$, i.e., P1.

Step 4: Proof of P2.

Using a similar argument, we can show that $\lim_{t\to\infty}\dot{\omega}_i(t) = 0$. It, together with (2.22), gives

$$\lim_{t\to\infty}\sum_{k\in\mathcal{N}_i(t)}\beta(\theta_{i,k}(t)) = 0.$$

From the fact of $\dot{\theta}_i = \omega_i$ in (2.22) and $\lim_{t\to\infty}\omega_i(t) = 0$ in the proof of Step 3, it can be deduced that $\lim_{t\to\infty}\dot{\theta}(t) = \mathbf{0}$. Moreover, as β is a continuous function of θ, one has $\lim_{t\to\infty}\theta(t) = \theta^*$ for θ^* satisfying

$$\sum_{k\in\mathcal{N}_i^*}\beta(\theta_{i,k}^*) = 0,\ \ \mathcal{N}_i^* = \{k \in \mathcal{V}, k \neq i \mid |\theta_{i,k}^*| < \gamma\}. \tag{2.36}$$

We introduce a simpler notation $\langle i, k\rangle := \theta_{i,k}^*$ to avoid using multiple layers of subscripts in the remaining proof.

To prove P2, it suffices to prove

$$|\langle i, k\rangle| \geq 2\pi/n,\ \ \forall i \neq k \in \mathcal{V}. \tag{2.37}$$

If all robots have no neighbor, it happens only when $\gamma = 2\pi/n$ and every two adjacent robots have the angle of γ, which proves (2.37) trivially. In the remaining proof, we consider the case that at least one robot has at least one neighbor. The details are given below.

For every $i \in \mathcal{V}$, k is called a front neighbor (FN) of i if $\langle k, i\rangle \in (0, \gamma)$, a rear neighbor (RN) of i if $\langle k, i\rangle \in (-\gamma, 0)$, and not a neighbor otherwise. Pick an arbitrary robot and label it as $s(0)$, and label the remaining robots in counter clockwise order as $s(1), s(2), \ldots, s(n-1)$. It is obvious that $\{s(0), s(1), s(2), \ldots, s(n-1)\} = \mathcal{V}$. For notation complement, we denote $s(i) = s(i \mod n)$ for any integer i, e.g., $s(n) = s(0)$.

Claim i: Every robot $i \in \mathcal{V}$ has at least one FN.

If this claim does not hold, there exists at least one robot that has no FN, say $s(i)$. In the sequence of $s(i), s(i-1), s(i-2), \ldots$, pick the first robot that has no FN but at least one RN, called $\hbar = s(i-\ell)$. Such a robot always exists because we consider the case that at least one robot has at least one neighbor. As a result,

$$\sum_{k\in\mathcal{N}_\hbar^*}\beta(\langle\hbar, k\rangle) = \sum_{k\in\mathcal{V},\langle\hbar,k\rangle\in(0,\gamma)}\beta(\langle\hbar, k\rangle) < 0,$$

which contradicts (2.36). The claim is proved.

Claim ii: For every $i \in \mathcal{V}$, $\langle s(i+1), s(i)\rangle \geq \gamma/2$.

If this claim does not hold, suppose $\langle s(0), s(n-1)\rangle < \gamma/2$ without loss of generality. Let $s(a_0), s(a_1), \ldots, s(a_m)$ be the longest sequence satisfying $a_0 = 0 < a_1 < \cdots < a_m < a_{m+1} = n$ and $\langle s(a_i), s(a_i - 1)\rangle < \gamma/2, i = 0, \ldots, m$. We consider two cases.

Case 1: If $\langle s(a_{i+1}), s(a_i)\rangle < \gamma/2$, there is no robot between $s(a_{i+1})$ and $s(a_i)$, that is, $a_{i+1} - a_i = 1$.

Case 2: If $\langle s(a_{i+1}), s(a_i)\rangle \geq \gamma/2$, then $a_{i+1} - a_i \geq 2$ and $\langle s(a_i + 1), s(a_i)\rangle > \gamma/2$. Due to $\langle s(a_i), s(a_i - 1)\rangle < \gamma/2$ and $\langle s(a_i + 1), s(a_i)\rangle > \gamma/2$, $s(a_i)$ has at least two FNs, that is, $\langle s(a_i + 2), s(a_i)\rangle < \gamma$. Otherwise,

$$\begin{aligned}\sum_{k \in N^*_{s(a_i)}} \beta(\langle s(a_i), k\rangle) \leq& \beta(\langle s(a_i), s(a_i + 1)\rangle) + \beta(\langle s(a_i), s(a_i - 1)\rangle)\\ =& -\beta(\langle s(a_i + 1), s(a_i)\rangle) + \beta(\langle s(a_i), s(a_i - 1)\rangle)\\ <& -\beta(\gamma/2) + \beta(\gamma/2) = 0,\end{aligned}$$

noting that $\beta(|s|)$ is a strictly increasing function of $|s|$ for $|s| \in (0, \gamma)$. It is a contradiction to (2.36). From $\langle s(a_i + 2), s(a_i)\rangle < \gamma$ and $\langle s(a_i + 1), s(a_i)\rangle > \gamma/2$, one has $\langle s(a_i + 2), s(a_i + 1)\rangle < \gamma/2$. As a result, $a_{i+1} - a_i = 2$ and $\langle s(a_{i+1}), s(a_i)\rangle < \gamma$.

The conclusions in the two cases can be unified as $\langle s(a_{i+1}), s(a_i)\rangle < (a_{i+1} - a_i)\gamma/2$. As a result,

$$2\pi = \sum_{i=0}^{m} \langle s(a_{i+1}), s(a_i)\rangle < \sum_{i=0}^{m} (a_{i+1} - a_i)\gamma/2 = n\gamma/2,$$

which is a contradiction to C3. Claim ii is thus proved.

Claim iii: Every robot $i \in \mathcal{V}$ has exactly one FN and one RN.

Every robot s_i has at least one FN, as stated in Claim i. However, it cannot have two FNs due to $\langle s(i+2) - s(i)\rangle = \langle s(i+2) - s(i+1)\rangle + \langle s(i+1) - s(i)\rangle \geq \gamma$ according to Claim ii. Consequently, each robot has exactly one FN, which also implies that each robot has exactly one RN.

Finally, according to Claim (iii), one has, for any s_i, for any robot, it holds that

$$\sum_{k \in N^*_{s(i)}} \beta(\langle s(i), k\rangle) = \beta(\langle s(i), s(i+1)\rangle) + \beta(\langle s(i), s(i-1)\rangle) = 0$$

according to (2.36), which further implies $\langle s(i+1), s(i)\rangle = \langle s(i), s(i-1)\rangle$ for all $i \in \mathcal{V}$, and hence $\langle s(i+1), s(i)\rangle = 2\pi/n$. This is equivalent to (2.37). Thus, property P2 is proven. □

Remark 2.4 The collision avoidance property P3 ensures that solutions such as $\rho_i = M$ and $\langle s(i+1), s(i)\rangle = 0$ for the conditions $\alpha(\rho_i(t)) = 0$ and $\sum_{k \in Ni(t)} \beta(\theta i, k(t)) = 0$ are not feasible. Furthermore, another solution where

$\langle s_{i+1}, s_i \rangle = \gamma$ for $i \in \mathcal{V}$ to $\sum_{k\in Ni(t)} \beta(\theta i, k(t)) = 0$ is ruled out due to the contradiction with $\sum_{i=0}^{n-1} \langle s(i+1), s(i) \rangle = 2\pi$ and condition C3. Thus, it follows that ρ_i and $\langle s(i+1), s(i) \rangle$ converge to the unique equilibrium point of ρ_o and $2\pi/n$, respectively, as time approaches infinity. Specifically, $\lim_{t\to\infty} \rho_i(t) = \rho_o$ and $\lim_{t\to\infty} \langle s(i+1), s(i) \rangle = 2\pi/n$, which results from $\lim_{t\to\infty} \alpha(\rho_i(t)) = 0$ and $\lim_{t\to\infty} \sum_{k\in Ni(t)} \beta(\theta i, k(t)) = 0$.

2.4 Simulation Results

In this section, we consider a setup with $n = 7$ robots using the structure defined by (2.1), (2.10), and (2.13), and a target described by (2.2) with an unknown constant velocity. The controller parameters are chosen as follows: $k_1 = 40,000$, $k_2 = 1.5$, $k_3 = 0.006$, $k_4 = 0.2$, $\rho_o = 30$, and $M = 20$. According to Remark 2.2, the observer (2.13) meets condition C1 with $\phi_1 = 1$ and $\phi_2 = 0.2$. The initial positions of the robots are set to $x_1(0) = [0, 30]^\top$, $x_2(0) = [0, 80]^\top$, $x_3(0) = [30, 0]^\top$, $x_4(0) = [0, 50]^\top$, $x_5(0) = [60, 0]^\top$, $x_6(0) = [10, 0]^\top$, $x_7(0) = [0, 10]^\top$, and the initial target position is $x_d(0) = [75, 85]^\top$, which satisfies condition C2. The initial velocities for both robots and the target are $v_i(0) = [0, 0]^\top$ for $i \in \mathcal{V}$, and $v_d = [0.7, 0.7]^\top$. The parameter $\gamma = 2.2\pi/7$ fulfills condition C3.

According to Theorem 2.1, the objective of fencing a moving target with a regular polygon formation while avoiding collisions is achieved if the properties P1, P2, and P3 are all guaranteed. Precisely, Fig. 2.1 presents the trajectories of the robots starting from various initial positions and the target with a simulation period of 100 s. Here, the dashed black polygon indicates the convex hull of the robots, and the red

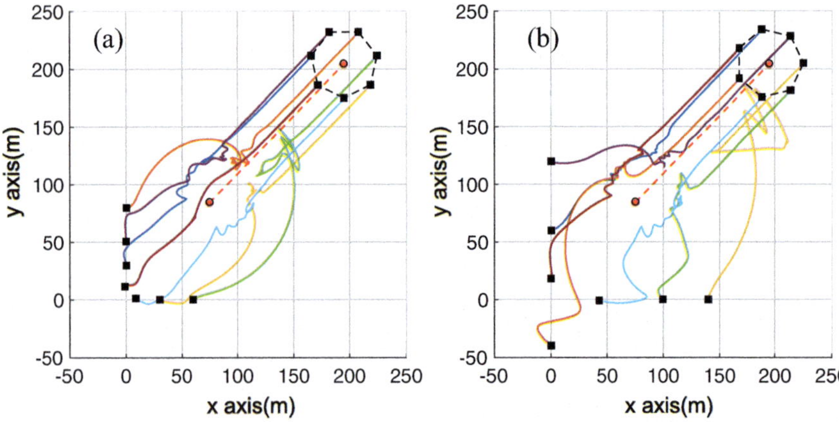

Fig. 2.1 **a, b** Moving-target-fencing trajectories of seven robots from different initial positions in a 2D plane. (Black rectangle: robot position with a solid line trajectory; red circle: target position with a dashed line trajectory)

circles represent the positions of the moving target. As shown in Fig. 2.1, a regular polygon is formed by the robots around the target. Then, we take Fig. 2.1a as an illustrative example, where the distance $\rho_i(t)$ between the robots and the target, along with the angle θ_i (adjusted using $+2\kappa\pi$ for an integer κ such that $\theta_i \in (-\pi, \pi]$ for clearer presentation), is shown in Figs. 2.2 and 2.3, respectively. These figures confirm properties P1 and P2, with $\lim_{t\to\infty} \rho_i(t) = 30$ and $\lim_{t\to\infty} |\theta_{i,k}(t)| \geq 2\pi/7$ for all $i \neq k \in \mathcal{V}$. Additionally, Figs. 2.2 and 2.3 demonstrate that $\rho_i(t) > 20$ and $\theta_{i,k}(t) \neq 0$ for all $i \neq k \in \mathcal{V}$ throughout the process $t \in [0, 100]$, validating the collision avoidance aspect of property P3. Finally, Figs. 2.2 and 2.3 illustrate the convergence

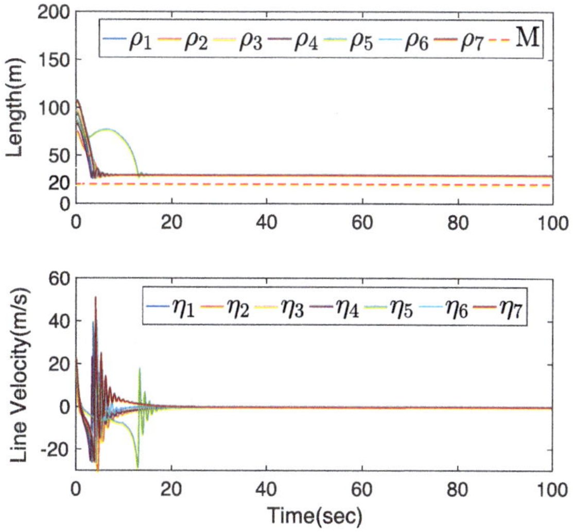

Fig. 2.2 Time evolution of the relative distance ρ_i and the line velocity η_i between each robot and the target in Fig. 2.1a for instance

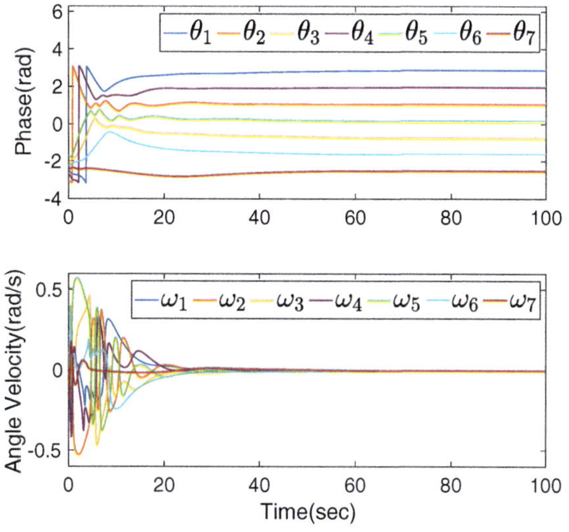

Fig. 2.3 Time evolution of the relative phase θ_i and the angle velocity ω_i between each robot and the target in Fig. 2.1a for instance

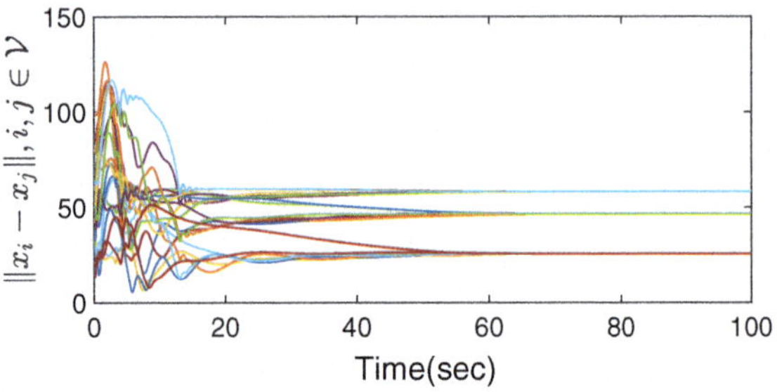

Fig. 2.4 Time evolution of the inter-robot distances $\|x_i - x_j\|$, $i, j \in \mathcal{V}$, in Fig. 2.1a for instance

of the linear velocity η_i and the angular velocity ω_i to zero for all $i \in \mathcal{V}$, as expected in the proof of Theorem 2.1. As shown in Fig. 2.4, the inter-robot distances satisfy $|x_i - x_j| > 0$ for all $i, j \in \mathcal{V}$, confirming that the robots successfully avoid collisions.

2.5 Conclusion

In this chapter, we have introduced a distributed fencing controller that enables a network of robots with second-order dynamics to enclose a moving target of unknown velocity within their convex hull. This approach establishes a solid foundation for more complex fencing missions to be explored in subsequent chapters. The proposed control scheme integrates three key functionalities: estimation of the target's velocity, regulation of the distance between the robots and the target, and angle repulsion among the robots. These ensure both the formation of a regular polygon and collision avoidance. The effectiveness of the control strategy has been demonstrated through simulation results. Future research could extend this method to accommodate more complex robot dynamics and practical scenarios such as obstacle avoidance. Potential applications of this algorithm include convoy protection using multiple unmanned surface vessels, dynamic target encirclement by unmanned systems, and collaborative monitoring of designated areas. These applications offer promising directions for further exploration.

References

1. J. Guo, G. Yan, Z. Lin, Local control strategy for moving-target enclosing under dynamically changing network topology. Syst. Control Lett. **59**(10), 654–661 (2010)
2. T.-H. Kim, T. Sugie, Cooperative control for target-capturing task based on a cyclic pursuit strategy. Automatica **43**(8), 1426–1431 (2007)
3. R. Zheng, Y. Liu, D. Sun, Enclosing a target by nonholonomic mobile robots with bearing-only measurements. Automatica **53**, 400–407 (2015)

4. R. Sharma, M. Kothari, C.N. Taylor, I. Postlethwaite, Cooperative target-capturing with inaccurate target information, in *Proceeding of the American Control Conference* (2010), pp. 5520–5525
5. X. Yu, L. Liu, Distributed circular formation control of ring-networked nonholonomic vehicles. Automatica **68**, 92–99 (2016)
6. X. Yu, L. Liu, Cooperative control for moving-target circular formation of nonholonomic vehicles. IEEE Trans. Autom. Control **62**(7), 3448–3454 (2016)
7. F. Chen, W. Ren, Y. Cao, Surrounding control in cooperative agent networks. Syst. Control Lett. **59**(11), 704–712 (2010)
8. Y.J. Shi, R. Li, K.L. Teo, Cooperative enclosing control for multiple moving targets by a group of agents. Int. J. Control **88**(1), 80–89 (2015)
9. S. Shoja, M. Baradarannia, F. Hashemzadeh, M. Badamchizadeh, P. Bagheri, Surrounding control of nonlinear multi-agent systems with non-identical agents. ISA Trans. **70**, 219–227 (2017)
10. R. Li, Y. Shi, Y. Song, Multi-group coordination control of multi-agent system based on smoothing estimator. IET Control Theory Appl. **10**(11), 1224–1230 (2016)
11. C. Li, L. Chen, Y. Guo, Y. Lyu, Cooperative surrounding control with collision avoidance for networked Lagrangian systems. J. Franklin Inst. **355**(12), 5182–5202 (2018)
12. B.-B. Hu, H.-T. Zhang, J. Wang, Multiple-target surrounding and collision avoidance with second-order nonlinear multiagent systems. IEEE Trans. Ind. Electron. **68**(8), 7454–7463 (2020)
13. C. Wang, G. Xie, Limit-cycle-based decoupled design of circle formation control with collision avoidance for anonymous agents in a plane. IEEE Trans. Autom. Control **62**(12), 6560–6567 (2017)
14. B. Liu, Z. Chen, H.-T. Zhang, X. Wang, T. Geng, H. Su, J. Zhao, Collective dynamics and control for multiple unmanned surface vessels. IEEE Trans. Control Syst. Technol. **28**(6), 2540–2547 (2019)
15. Z. Chen, A cooperative target-fencing protocol of multiple vehicles. Automatica **107**, 591–594 (2019)
16. B.A. Francis, W.M. Wonham, The internal model principle of control theory. Automatica **12**(5), 457–465 (1976)
17. Z. Chen, J. Huang, *Stabilization and regulation of nonlinear systems* (Springer, Cham, Switzerland, 2015)
18. R. Freeman, Global internal stabilizability does not imply global external stabilizability for small sensor disturbances. IEEE Trans. Autom. Control **40**(12), 2119–2122 (1995)
19. K.H. Khalil, *Nonlinear Systems* (Prentice Hall, Upper Saddle River, 2002)

Chapter 3
Distributed Fencing Control via Input-to-State Stability

3.1 Introduction

In Chap. 2, we introduced a distributed fencing controller via neighboring angle repulsion to achieve cooperative fencing of a constant-speed target, establishing the fundamental task of fencing. This chapter considers the more complex fencing problem that ensures evenly rotational formation during the fencing process, which is distinct from the previous fencing work in Chap. 2. Moreover, it also considers the disturbances induced by intermittent topologies during the fencing process. A distributed cooperative fencing control algorithm via input-to-state stability is proposed, which aims to solve the more complex fencing problem with evenly rotational formation. In what follows, we will introduce the related background.

Over the years, coordinated control technologies for multiple USVs have advanced significantly, driven by the growing demands for aquatic resource exploration, rescue operations, and waterway logistics. Compared to single USVs, multi-USV systems offer notable benefits, including higher efficiency, greater flexibility, adaptability to challenging environments, wider operational coverage, and lower maintenance costs. These features are crucial for undertaking complex tasks such as ocean resource exploration, water quality monitoring, and maritime logistics. Despite these advantages, much of the existing research has focused on single-vessel operations, addressing challenges like path planning and obstacle avoidance [1–3], trajectory tracking [4–6], and stabilization and anchoring [7]. However, single USVs are limited in both capability and coverage, making them unsuitable for more demanding tasks such as coordinated patrolling, rescue missions, and exploration. This highlights the critical need for the development of distributed cooperative control strategies specifically designed for multi-USV systems.

In this endeavor, some initial efforts have been devoted to the formation strategies of multi-USV systems, such as the leader–follower frameworks [8–10]. While these methods are effective, they rely heavily on leader information to achieve shared objectives, making them less adaptable to environmental uncertainties and more

B.-B. Hu and H.-T. Zhang, *Multi-Robot Systems*, Cognitive Intelligence and Robotics,
https://doi.org/10.1007/978-981-97-9214-6_3

vulnerable to external disturbances. Additionally, efforts have been made in collective formation control, where USVs are coordinated to form desired geometric patterns while following prescribed trajectories [11–13]. Due to the complexity of designing interaction protocols for specific formations, only a limited number of studies have focused on containment control, in which USVs are guided to a convex hull surrounding multiple targets or leaders [14, 15].

To accomplish the collective protection of a specified target, effective fencing control strategies are required for multi-agent systems (MASs). Some studies have addressed scenarios where all agents have complete knowledge of the target's states. Yamaguchi [16] introduced a feedback control law that allows multiple robots to fence in a target robot. Ni and Yang [17] proposed a neural network-based control approach for collective pursuit and capture. Sepulchre et al. [18] developed distributed control protocols to organize unicycles into a circular formation using all-to-all communication. Marshall et al. [19] extended circular fencing formation control for practical task execution. Sinha and Ghose [20] designed a circulation formation control strategy tailored for heterogeneous MASs. Kim and Sugie [21] developed a cyclic controller to achieve the desired pattern for a moving target in 3D space. Ceccarelli and Macro et al. [22] introduced a cyclic pursuit strategy considering the limited visibility of onboard sensors. Li et al. [23] presented a cooperative circumnavigation control protocol that guides a group of micro-satellites to maintain a planar elliptical orbit with a specified radius.

In light of the significant maintenance costs involved in real-world applications, there has been a shift in research toward MAS control approaches where only a subset of vessels have information about the target(s). For instance, a hybrid control scheme for full cyclic formation was proposed in [24], while [25] introduced a distributed cooperative control law where just one agent is informed of the target's location. In [26], a cycle formation control strategy based solely on bearing measurements was presented. Moreover, a global-orientation-free approach was developed in [27] to guide MASs in circumnavigating an unknown target in 3D space. To tackle scenarios involving dynamically switching topologies, [28] proposed a distributed control protocol for hunting a moving target, which was later refined into a more practical distributed dynamic controller in [29] that operates using only intermittent neighbor information.

Although most existing MAS fencing studies focus on scenarios where target information is fully accessible to all agents [16–23], the substantial maintenance costs limit their broader application, particularly in complex maritime operations. A few studies have explored MAS fencing control where only a limited number of agents have access to the target [24–29]. However, these approaches mainly address static targets using agents with first-order or second-order dynamics. The challenge of developing a fencing control strategy for more practical moving targets, especially involving underactuated USVs, remains unresolved. Furthermore, intermittent communications among USVs, which is common in real-world applications, pose additional challenges. Experimental studies in multi-USV fencing control are still scarce, though they are essential for bridging the gap between theoretical models and real-world deployment. As a result, it is both crucial and demanding to create an advanced distributed control protocol for multi-USV systems to effectively encircle a moving target.

In this chapter, we present a distributed fencing control framework tested on an actual multi-USV setup, which includes three HUSTER-12 USVs, each 1.6 m in length, along with a differential GPS station and a dock at Songshan Lake in Dongguan City, Guangdong Province, China. The framework features a two-stage control strategy for multi-USV systems, where only a subset of the vessels has information about the moving target. An observer is utilized to estimate the target's state. In the first stage, the target is enclosed by a convex hull formed by the USVs positioned at the vertices. In the second stage, the vessels capture the target and rotate uniformly around it at its geographical center. Experimental results confirm the effectiveness of this approach. The main contributions of this chapter are two-fold.

1. Propose a two-stage fencing controller for multi-USV systems with underactuated dynamics that rely solely on local information and intermittent inter-USV communications.
2. Demonstrate the effectiveness of the proposed control framework with a practical experiment involving a setup of three 1.6-meter-long USVs and a differential GPS station at a dock on a lake.

The remainder of this chapter is structured as follows: Sect. 3.2 presents the formulation of the fencing control problem for USVs and introduces the proposed two-stage fencing control framework. In Sects. 3.3 and 3.4, an estimator for the moving target is designed, and a two-stage distributed fencing control protocol is developed. Stability conditions are derived to ensure the proposed controller's functionality. Section 3.5 includes experimental results to validate the effectiveness of the control framework. Finally, Sect. 3.6 provides the conclusions.

3.2 Problem Formulation of Fencing with an Evenly Rotating Formation

In this section, we consider a multi-USV system with n vessels, where the positions are given by $q := [q_1^\top, q_2^\top, \ldots, q_n^\top]^\top \in \mathbb{R}^{2n}$, with $q_i = [x_i, y_i]^\top \in \mathbb{R}^2$ representing the 2D coordinates of vessel i. The communication network among the USVs is defined by the graph $\mathcal{G}(\mathcal{V}, \mathcal{E})$, where $\mathcal{V} = \{1, \ldots, n\}$ denotes the set of vessels, and ε represents the edges connecting them. Let $\mathcal{N}_i$ be the neighborhood of vessel i within this graph. The symmetric adjacency matrix $A = [a_{ij}] \in \mathbb{R}^{n\times n}$ characterizes the undirected communication graph, where $a_{ij} = a_{ji}$. Specifically, $a_{ij} = 1$ if there is an edge between vessels i and j, and $a_{ij} = 0$ otherwise. Self-loops are not allowed, hence $a_{ii} = 0$. The connection matrix $B(t) = \mathrm{diag}\{b_i\} \in \mathbb{R}^{n\times n}$ defines the interaction between the vessels and the target, with $b_i = 1$ if vessel i detects the target q_0, and $b_i = 0$ otherwise.

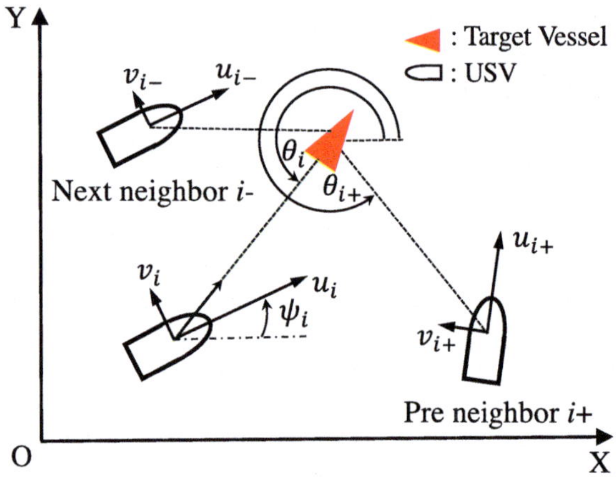

Fig. 3.1 Illustration of USV velocities and neighbors

Now, we introduce a dynamic model for the multi-USV system and proceed to describe the concept of fencing control. First of all, the kinetic model of USV i is

$$\dot{q}_i = J(\psi_i)\mathbf{v}_i, \tag{3.1}$$

with

$$J(\psi_i) = [s_{\psi_i}, t_{\psi_i}], s_{\psi_i} = \begin{bmatrix} \cos(\psi_i) \\ \sin(\psi_i) \end{bmatrix}, t_{\psi_i} = \begin{bmatrix} -\sin(\psi_i) \\ \cos(\psi_i) \end{bmatrix}. \tag{3.2}$$

According to Fig. 3.1, ψ_i represents the phase angle of vessel i relative to the earth frame, and $\mathbf{v}_i = [u_i, v_i]^\top \in \mathbb{R}^2$ is the velocity vector in the body-fixed coordinate system, where u_i denotes the surge velocity and v_i denotes the sway velocity. The neighbor relationships in the multi-USV system are established based on their relative positions in a clockwise sequence around the target, as depicted in Fig. 3.1 as well. Therefore, vessel i's pre-neighbor is labeled i^+, while its next-neighbor is labeled i^-.

Then, we introduce the dynamics model of each vessel (see, e.g., [4])

$$\begin{aligned}
\dot{\psi}_i &= r_i, \\
\dot{u}_i &= \frac{m_{22}}{m_{11}} v_i r_i - \frac{d_{11}}{m_{11}} u_i + \frac{\tau_{iu}}{m_{11}}, \\
\dot{r}_i &= \frac{m_{11} - m_{22}}{m_{33}} u_i v_i - \frac{d_{33}}{m_{33}} r_i + \frac{\tau_{ir}}{m_{33}}, \\
\dot{v}_i &= -\frac{m_{11}}{m_{22}} u_i r_i - \frac{d_{22}}{m_{22}} v_i,
\end{aligned} \tag{3.3}$$

where $i \in \mathcal{V}$, m_{ii} and d_{ii} are the mass and damping coefficients, respectively, $\boldsymbol{\tau}_i = [\tau_{iu}, \tau_{ir}]^\top$ denotes the control inputs, and r_i is the rate of yaw.

Let ψ_i^r, u_i^r, and v_i^r represent the desired values for ψ_i, u_i, and v_i, respectively. Define

$$\widehat{\psi}_i := \psi_i - \psi_i^r, \widehat{u}_i := u_i - u_i^r, \widehat{v}_i := v_i - v_i^r. \tag{3.4}$$

Substituting (3.4) into (3.1) yields

$$\dot{q}_i = \mathbf{u}_i^r + e_i, \tag{3.5}$$

with

$$\begin{aligned} \mathbf{u}_i^r &:= J(\psi_i^r)\begin{bmatrix} u_i^r \\ v_i^r \end{bmatrix}, \\ e_i &:= [J(\psi_i^r + \widehat{\psi}_i) - J(\psi_i^r)]\begin{bmatrix} u_i^r \\ v_i^r \end{bmatrix} + J(\psi_i)\begin{bmatrix} \widehat{u}_i \\ \widehat{v}_i \end{bmatrix}. \end{aligned} \tag{3.6}$$

Let $\mathrm{co}(\mathcal{V})$ be the convex hull of $q_1, q_2, \ldots, q_n$, i.e.,

$$\mathrm{co}(\mathcal{V}) := \left\{ \sum_{\mathrm{i=1}}^{\mathrm{n}} \lambda_{\mathrm{i}} \mathrm{q}_{\mathrm{i}} \middle| \lambda_{\mathrm{i}} \geq 0, \sum_{\mathrm{i=1}}^{\mathrm{n}} \lambda_{\mathrm{i}} = 1 \right\}. \tag{3.7}$$

Define $D_{q_0}(p) := \min_{p \in \mathrm{co}(\mathcal{V})} |q_0(t) - p|$ as the distance from the target to the convex hull $\mathrm{co}(\mathcal{V})$, where $q_0(t) := [x_0(t), y_0(t)]^\top$ represents the target's position (see, e.g., [30]).

Next, we define fencing control and equal fencing control as follows.

Definition 3.1 ([30] *Fencing*) The multi-USV system $\mathcal{G}(\mathcal{V}, \mathcal{E})$ with individual dynamics described by (3.5) achieves fencing status if all the USVs are positioned such that they form a convex hull enclosing the target vessel, i.e.,

$$\lim_{t \to \infty} D_{q_0}(t) := \min_{p \in \mathrm{co}(\mathcal{V})} \|q_0(t) - p(t)\| = 0, \tag{3.8}$$

where $\mathrm{co}(\mathcal{V})$ is given in (3.7).

Definition 3.2 ([30] *Equal Fencing*) The multi-USV system $\mathcal{G}(\mathcal{V}, \mathcal{E})$ with individual dynamics given by (3.5) achieves equal fencing when the fencing status is fulfilled and all the USVs rotate evenly around the target vessel, i.e.,

$$\begin{aligned} &\lim_{t \to \infty} \rho_i := \|q_i - q_0\| = r, \\ &\lim_{t \to \infty} \theta_{i+} - \theta_i + \xi_i = \lim_{t \to \infty} \theta_{j+} - \theta_j + \xi_j, \end{aligned} \tag{3.9}$$

with

$$\xi_i := \begin{cases} 0, & \theta_{i+} - \theta_i \geq 0, \\ 2\pi, & \theta_{i+} - \theta_i < 0, \end{cases} \tag{3.10}$$

and $i, j \in \mathcal{V}, i \neq j$. Here, r represents the radius of encirclement, and $\theta_i \in [0, 2\pi)$ denotes the relative angle of vessel i with respect to the target q_0, as illustrated in Fig. 3.1.

So far, many multi-USV control strategies have been based on the assumption of a consistently connected communication network. In practice, however, factors such as waves, tides, and winds can disrupt inter-vessel communication, leading to periods of intermittent connectivity. To tackle this challenge, we introduce the definition of jointly dwelling connection.

Definition 3.3 ([31] *Jointly dwelling connection*) The condition is met if $T_c \geq \mu T$ for a given rate $0 < \mu < 1$ and $T > 0$. Here, T_c represents the minimum duration of time intervals $[T_j, T_{j+1})$, where $T_0 = t_0$ and $T_{j+1} = T_j + T$. Within each interval $[T_j, T_{j+1})$, T_j^c denotes the total duration during which the communication graphs are connected. These intervals are composed of several subintervals $[t_j, t_{j+1})$, where the interconnection graph remains constant. It is assumed that each subinterval satisfies $t_{j+1} - t_j \geq \tau$, with τ defined as the dwell time.

We are now ready to present three technical problems addressed in this paper.

Problem 3.1 (*Fencing Control*) Develop a control protocol

$$\mathbf{u}_i^r = f(q_i, q_j), \quad i \in \mathcal{V}, j \in \mathcal{N}_i, \tag{3.11}$$

for vessel i in a multi-USV system $\mathcal{G}(\mathcal{V}, \mathcal{E})$ described by (3.5). The objective is to achieve fencing (as shown in Definition 3.1) around a moving target $q_0(t) \in \mathbb{R}^2$ with a constant velocity $\dot{q}_0(t) = v_0$, ensuring that the error $e_i(t)$ converges to zero as t approaches infinity.

Problem 3.2 (*Equally Fencing Control*) Design a control protocol

$$\mathbf{u}_i^r = f(q_i, q_j), \quad i \in \mathcal{V}, j \in \mathcal{N}_i, \tag{3.12}$$

for vessel i in a multi-USV system $\mathcal{G}(\mathcal{V}, \mathcal{E})$ governed by (3.5). The objective is to achieve an equal fencing configuration with an evenly rotating formation (as defined in Definition 3.2) around a target $q_0(t) \in \mathbb{R}^2$ moving with a constant velocity $\dot{q}_0(t) = v_0$, ensuring that the error $e_i(t)$ converges to zero as t approaches infinity.

For the condition of $v_0 = 0$, Problems 3.1 and 3.2 specifically address fencing control and equally fencing control for a stationary target, respectively. To solve the overarching collective fencing control problem, each vessel must be regulated to achieve the target velocities ψ_i^r, u_i^r, v_i^r.

Problem 3.3 (*Individual Regulation*) Develop a control protocol $\boldsymbol{\tau} = (\tau_{iu}, \tau_{ir})^\top$ for each vessel in the multi-USV system described by (3.3), ensuring that $\lim_{t\to\infty} \psi_i = \psi_i^r$, $\lim_{t\to\infty} u_i = u_i^r$, $\lim_{t\to\infty} v_i = v_i^r$, and $\lim_{t\to\infty} e_i = \mathbf{0}_2$.

Since Problem 3.3 has been previously solved in our earlier research [30], this study will focus on addressing the high-level collective fencing control challenges posed by Problems 3.1 and 3.2, specifically finding the desired control signals $\mathbf{u}_i^r := [\psi_i^r, u_i^r, v_i^r]^\top$ for each vessel in the multi-USV system.

3.3 Distributed Target Estimation Under Intermittent Topologies

To address Problem 3.1, we need to develop a distributed estimator for each vessel that can estimate the state of the moving target, given its constant velocity $\dot{q}_0(t) = v_0 \in \mathbb{R}^2$, using intermittent information from neighboring vessels. The estimated states, $\widehat{q}_0^i$ and $\widehat{v}_0^i$, will then be used to design the feedback control $\mathbf{u}_i^r$ for vessel i.

The distributed controller is depicted in Fig. 3.2. In Step 1, the control inputs $\mathbf{u}_i^r$, which include the desired velocities u_i^r and v_i^r and the heading ψ_i^r, are determined using $f(q_i, q_j)$ and the estimated target position $\widehat{q}_0^i$. In Step 2, an individual control law is developed to address Problem 3.3. To continue, the following two assumptions must be established.

Assumption 3.1 Initially, all USVs in the system $\mathcal{G}(\mathcal{V}, \mathcal{E})$ are positioned within a circle of sensing radius R, meaning that for any pair of vessels i and j in $\mathcal{V}$, the distance between them satisfies $|q_i(0) - q_j(0)| \leq 2R$.

Assumption 3.2 In the multi-USV system $\mathcal{G}(\mathcal{V}, \mathcal{E})$, there is always at least one vessel that has access to the target q_0, meaning that the communication matrix $B(t)$ always contains at least one entry $b_i(t) \neq 0$.

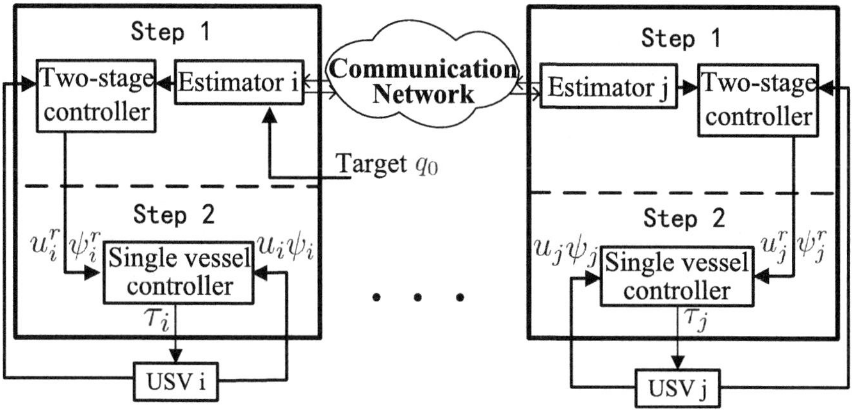

Fig. 3.2 Two-level fencing controller structure of the multi-USV system, where USV i has access to the target q_0 but USV j does not

Let $\widehat{q}_0^i(t)$ and $\widehat{v}_0^i(t)$ represent the estimated position and velocity of the target $q_0(t)$, respectively, as determined by USV i. These estimates are obtained following the methodology outlined in [31], where they are given by

$$\begin{aligned}\dot{\widehat{q}}_0^i(t) = & -k_1[\sum_{j\in N_i}(\widehat{q}_0^i(t)-\widehat{q}_0^j(t))+b_i(t)(\widehat{q}_0^i(t)-q_0(t))]+\widehat{v}_0^i(t),\\ \dot{\widehat{v}}_0^i(t) = & -\gamma k_1[\sum_{j\in N_i}(\widehat{q}_0^i(t)-\widehat{q}_0^j(t))+b_i(t)(\widehat{q}_0^i(t)-q_0(t))].\end{aligned} \tag{3.13}$$

Let $\widehat{q}_0(t) := [\widehat{q}_0^1(t)^\top, \widehat{q}_0^2(t)^\top, \ldots, \widehat{q}_0^n(t)^\top]^\top, \widehat{v}_0(t) := [\widehat{v}_0^1(t)^\top, \widehat{v}_0^2(t)^\top, \ldots, \widehat{v}_0^n(t)^\top]^\top \in \mathbb{R}^{2n}$, and rewrite (3.13) in a compact form as

$$\begin{aligned}\dot{\widehat{q}}_0(t) = & -k_1(L_G(t)+B(t))\otimes I_2\widehat{q}_0+(k_1B(t)\otimes I_2)\mathbf{1}_n\otimes q_0+\widehat{v}_0,\\ \dot{\widehat{v}}_0(t) = & -\gamma k_1(L_G(t)+B(t))\otimes I_2\widehat{q}_0+(\gamma k_1B(t)\otimes I_2)\mathbf{1}_n\otimes q_0.\end{aligned} \tag{3.14}$$

Lemma 3.1 ([31]) *For a first-order MAS with a moving target $q_0(t)$ and estimator governed by (3.14) with $0 < \gamma < 1$ and*

$$k_1 \geq \frac{1}{4\gamma(1-\gamma^2)\overline{\lambda}}, \tag{3.15}$$

where $\overline{\lambda} = \min\{\lambda(L_G + B(t))$, if Assumption 3.2 holds and $\mathcal{G}(\mathcal{V}, \mathcal{E})$ is jointly dwelling connected (see Definition 3.3), then the error of estimators converges to zero:

$$\lim_{t\to\infty}\widetilde{q}_0^i(t) = \mathbf{0}_4, \lim_{t\to\infty}\widetilde{v}_0^i(t) = \mathbf{0}_4 \tag{3.16}$$

with $\widetilde{q}_0^i := \widehat{q}_0^i - q_0$ and $\widetilde{v}_0^i := \widehat{v}_0^i - v_0$.

Based on Lemma 3.1, the estimates $\widehat{q}_0^i(t)$ and $\widehat{v}_0^i(t)$ for the target's position and velocity, respectively, will converge to the actual values of $q_0(t)$ and v_0, even when the network $\mathcal{G}(\mathcal{V}, \mathcal{E})$ experiences causal disconnections. This enhancement in robustness addresses communication failures more effectively compared to previous work such as [30].

3.4 Controller Design and Convergence Analysis

3.4.1 Target Fencing Control—Stage 1

According to the estimates $\widehat{q}_0^i$ and $\widehat{v}_0^i$ in (3.14), the objective is to design a control law $\mathbf{u}_i^r = f(q_i, q_j, \widehat{q}_0^i, \widehat{v}_0^i, \mathcal{G}), i, j \in \mathcal{V}$, that guides vessel i into a convex hull $\text{co}(\mathcal{V})$ surrounding the moving target $q_0(t)$, while ensuring $\lim_{t\to\infty} e_i(t) = 0$. To achieve this, a distributed control law $\mathbf{u}_i^r$ is proposed as follows:

$$\mathbf{u}_i^r = \widehat{q}_0^i + \delta_i - q_i + \widehat{v}_0^i, \tag{3.17}$$

where $\delta_i := \delta[\cos(2\pi i/n + \theta_0), \sin(2\pi i/n + \theta_0)]^\top$, and $\delta = \max |q_i(0) - q_j(0)|$, $i, j \in \mathcal{V}$.

Now, we present the main technical result related to Problem 3.1.

Theorem 3.1 *Under the system dynamics described by Eqs. (3.5), (3.13), and (3.17), and given Assumptions 3.1 and 3.2, Problem 3.1 for a moving target with* $\dot{q}_0(t) = v_0$ *is resolved if* $\lim_{t\to\infty} e_i(t) = 0$ *and the network* $\mathcal{G}(\mathcal{V}, \mathcal{E})$ *is jointly dwellingly connected.*

Proof Let $z_i := q_i - q_0 - \delta_i$. Then, one has

$$\dot{z}_i = \dot{q}_i - v_0. \tag{3.18}$$

Substituting (3.5) into (3.18) yields

$$\dot{z}_i = \mathbf{u}_i^r + e_i - v_0. \tag{3.19}$$

Then, the closed-loop system (3.19) can be written as

$$\dot{z}_i = \widehat{q}_0^i + \delta_i - q_i + \widehat{v}_0^i + e_i - v_0 = q_0 + \delta_i - q_i + \widetilde{e}_i = -z_i + \widetilde{e}_i, \tag{3.20}$$

where $\widetilde{e}_i = \widetilde{q}_0^i + \widetilde{v}_0^i + e_i$ and $\widetilde{q}_0^i := \widehat{q}_0^i - q_0$, $\widetilde{v}_0^i := \widehat{v}_0^i - v_0$ are estimated errors. It follows from Lemma 3.1 that $\lim_{t\to\infty}[\widetilde{q}_0^i(t)^\top, \widetilde{v}_0^i(t)^\top]^\top = \mathbf{0}_4$, which implies that $\lim_{t\to\infty} \widetilde{e}_i(t) = \mathbf{0}_2$. Rewrite (3.20) in a compact form as

$$\dot{z} = -z + \widetilde{e}, \tag{3.21}$$

with $z = [z_1, z_2, \ldots, z_n]^\top$ and $\widetilde{e} = [\widetilde{e}_1, \widetilde{e}_2, \ldots, \widetilde{e}_n]^\top \in \mathbb{R}^n$. Pick a Lyapunov function $V(z) = \frac{1}{2}z^\top z$, satisfying

$$\begin{aligned} \dot{V}(z) =& -z^\top z + z^\top \widetilde{e} \\ \leq& -\|z\|^2 + \|z\| \cdot \|\widetilde{e}\| \\ \leq& -(1 - c_1)\|z\|^2 + \|z\|(\|\widetilde{e}\| - c_1\|z\|), \end{aligned} \tag{3.22}$$

with $c_1 \in (0, 1)$. Then, one has $\dot{V}(z) \leq -(1 - c_1)\|z\|^2$, $\forall \|z\| \geq \|\widetilde{e}\|/c_1$. If $\|z(0)\| \geq \|\widetilde{e}\|/c_1$, then $\dot{V}(z) \leq 0$, which implies that $\|z(t)\|$ converges into the ball of radius $\|\widetilde{e}\|/c_1$. If $\|z(0)\| < \|\widetilde{e}\|/c_1$, it is clear that $\|z(0)\|$ has already been in the ball of radius $\|\widetilde{e}\|/c_1$. Both cases satisfy the definition of input-to-state stability (ISS) [32], which implies that $\|z(0)\|$ could be randomly picked. Accordingly, the closed-loop system (3.21) is ISS with respect to $\widetilde{e}$ that satisfies

$$\|z(t)\| \leq \beta(\|z(0\|, t)) + \gamma\big(\sup_{0\leq\tau\leq t} \|\widetilde{e}(\tau)\|\big), \forall t > 0, \tag{3.23}$$

where $\beta(\cdot)$ is a $\mathcal{KL}$ function and $\kappa_1 = 1/c_1$ is the gain of the system (see [33]). Due to $\lim_{t\to\infty} \widetilde{e}_i(t) = 0$ and the $\mathcal{KL}$ function satisfying $\lim_{t\to\infty} \beta(\|z(0\|, t)) = 0$, it follows that $\lim_{t\to\infty} z_i(t) = 0$ in (3.23). Moreover, one has

$$\lim_{t\to\infty} \sum_{i=1}^{n} \frac{1}{n} q_i = \lim_{t\to\infty} \frac{1}{n} \sum_{i=1}^{n} (z_i + q_0 + \delta_i) = \lim_{t\to\infty} \sum_{i=1}^{n} \frac{1}{n} q_0 = q_0, \tag{3.24}$$

which implies that the target q_0 is enclosed by n USVs within a convex hull $co(\mathcal{V})$, thereby solving Problem 3.1. This completes the proof. □

Remark 3.1 The estimator $\widehat{q}_0^i(t)$ and $\widehat{v}_0^i(t)$ in (3.13) will converge to the moving target $\dot{q}_0(t) = v_0$ along with the target's position $q_0(t)$ if the connection network $\mathcal{G}(\mathcal{V}, \mathcal{E})$ is jointly dwelling connected, i.e., $T_c \geq \mu T$. Otherwise, the estimator $\widehat{q}_0^i(t)$ and $\widehat{v}_0^i(t)$ will be unstable. It is important to note that each leader USV (a USV with potential access to the target based on the communication topology) transmits its estimated leader states (velocity and position information) to its neighbors, though this information is not necessarily exact. This communication is continuous and sufficient for controller design. As a result, the convergence of the target state estimator and the asymptotic stability of the closed-loop system can be achieved.

3.4.2 *Equal Fencing with Evenly Rotating Formation—Stage 2*

Problem 3.1 is resolved in Stage 1 for the moving target $q_0(t)$, where all vessels form a convex hull surrounding the target $q_0(t)$. Next, to address Problem 3.2, a second-stage control law $\mathbf{u}_i^r = f(q_i, q_j, \widehat{q}_0^i, \widehat{v}_0^i, \mathcal{G})$ is introduced as follows:

$$\mathbf{u}_i^r = k_2(r - \widehat{\rho}_i)s_{\widehat{\theta}_i} + \widehat{\rho}_i(\widehat{\theta}_{i+} - \widehat{\theta}_i + \xi_i)t_{\widehat{\theta}_i} + \widehat{v}_0^i, \tag{3.25}$$

where the parameter $k_2 > 0$, and the values of r and ξ_i are specified in Definition 3.1. Additionally, $\widehat{\rho}_i := |q_i - \widehat{q}_0^i|$ represents the distance between vessel i and the estimated target $\widehat{q}_0^i$, while $\widehat{\theta}_i$ denotes the relative angle from vessel i to the estimated target $\widehat{q}_0^i$. The angle θ_{i+} is accessible since the target is already encircled by the USVs, allowing each vessel to detect its next-neighbor as illustrated in Fig. 3.1.

The main technical result related to Problem 3.2 is established as follows.

Theorem 3.2 *For the system governed by Eqs. (3.5) and (3.25), Problem 3.2 for a moving target $\dot{q}_0(t) = v_0$ is solved, i.e.,*

$$(i) \lim_{t\to\infty} \rho_i = r, i = 1, 2, \ldots, n;$$
$$(ii) \lim_{t\to\infty} \theta_{i+} - \theta_i + \xi_i = \lim_{t\to\infty} \theta_{j+} - \theta_j + \xi_j, i, j \in \mathcal{V}, i \neq j,$$

if $\lim_{t\to\infty} e_i(t) = 0$ and the network $\mathcal{G}(t)$ is jointly dwellingly connected.

Proof According to the definitions of ρ_i, θ_i in Fig. 3.1, one has

$$q_i = \rho_i s_{\theta_i} + q_0. \tag{3.26}$$

Meanwhile, it follows from (3.18) that

$$\dot{z}_i = \dot{\rho}_i s_{\theta_i} + \rho_i t_{\theta_i} \dot{\theta}_i = J(\theta_i) \begin{bmatrix} \dot{\rho}_i \\ \rho_i \dot{\theta}_i \end{bmatrix}. \tag{3.27}$$

Since q_i can also be rewritten as $q_i = \widehat{\rho}_i s_{\widehat{\theta}_i} + \widehat{q}_0^i$, which implies

$$\lim_{t\to\infty} \widehat{\rho}_i = \rho_i, \ \lim_{t\to\infty} \widehat{\theta}_i = \theta_i, \lim_{t\to\infty} \widehat{\theta}_{i+} = \theta_{i+},$$

if $\lim_{t\to\infty} \widehat{q}_0^i(t) = q_0(t)$. Substituting (3.25) into the closed-loop system (3.19) yields

$$\begin{aligned} \dot{z}_i &= \mathbf{u}_i^r + e_i \\ &= k_2(r - \rho_i + \rho_i - \widehat{\rho}_i)(s_{\theta_i} + s_{\widehat{\theta}_i} - s_{\theta_i}) + (\rho_i + \widehat{\rho}_i \\ &\quad - \rho_i)(\theta_{i+} - \theta_i + \xi_i + \widehat{\theta}_{i+} - \widehat{\theta}_i - \theta_{i+} + \theta_i)(t_{\theta_i} + t_{\widehat{\theta}_i} - t_{\theta_i}) + \widehat{v}_0^i + e_i - v_0 \\ &= k_2(r - \rho_i)s_{\theta_i} + \rho_i(\theta_{i+} - \theta_i + \xi_i)t_{\theta_i} + \widetilde{e}_i, \end{aligned} \tag{3.28}$$

where $\widetilde{e}_i := k_2(r - \rho_i)(s_{\widehat{\theta}_i} - s_{\theta_i}) + k_2(\rho_i - \widehat{\rho}_i)s_{\widehat{\theta}_i} + \rho_i(\widehat{\theta}_{i+} - \widehat{\theta}_i - \theta_{i+} + \theta_i)t_{\widehat{\theta}_i} + (\widehat{\rho}_i - \rho_i)(\widehat{\theta}_{i+} - \widehat{\theta}_i + \xi_i)t_{\widehat{\theta}_i} + \widehat{v}_0^i + e_i - v_0$. From Lemma 3.1, it follows that $\lim_{t\to\infty} \widehat{q}_0^i(t) = q_0(t)$, $\lim_{t\to\infty} \widehat{v}_0^i(t) = v_0$, which implies that $\lim_{t\to\infty} s_{\widehat{\theta}_i}(t) - s_{\theta_i}(t) = 0$, $\lim_{t\to\infty} \rho_i(t) - \widehat{\rho}_i(t) = 0$, $\lim_{t\to\infty} \widehat{\theta}_i(t) - \theta_i(t) = 0$, and $\lim_{t\to\infty} \widehat{\theta}_{i+}(t) - \theta_{i+}(t) = 0$. Therefore, one has $\lim_{t\to\infty} \widetilde{e}_i(t) = 0$. Analogously, denoting

$$\widetilde{e}_i := J(\theta_i) \begin{bmatrix} \widehat{\eta}_i \\ \rho_i \widehat{\omega}_i \end{bmatrix}, \tag{3.29}$$

with $\lim_{t\to\infty} \widehat{\eta}_i(t) = 0$, $\lim_{t\to\infty} \widehat{\omega}_i(t) = 0$ (see, e.g., [30]), it follows from Eqs. (3.19), (3.25), and (3.29) that

$$z_i = J(\theta_i) \begin{bmatrix} k_2(r - \rho_i) + \widehat{\eta}_i \\ \rho_i(\theta_{i+} - \theta_i + \xi_i) + \rho_i \widehat{\omega}_i \end{bmatrix}. \tag{3.30}$$

Besides, from Eqs. (3.27) and (3.28), one has

$$\begin{cases} \dot{\rho}_i = k_2(r - \rho_i) + \widehat{\eta}_i, \\ \dot{\theta}_i = \theta_{i+} - \theta_i + \xi_i + \widehat{\omega}_i. \end{cases} \tag{3.31}$$

Define the equally cyclic fencing errors as

$$\begin{cases} \widetilde{\rho}_i := r - \rho_i, \\ \widetilde{\theta}_i := \frac{2\pi}{n} - (\theta_{i+} - \theta_i + \xi_i), \end{cases} \tag{3.32}$$

with

$$\rho = [\rho_1, \rho_2, \ldots, \rho_n]^{\mathsf{T}}, \widetilde{\rho} = [\widetilde{\rho}_1, \widetilde{\rho}_2, \ldots, \widetilde{\rho}_n]^{\mathsf{T}}, \theta = [\theta_1, \theta_2, \ldots, \theta_n]^{\mathsf{T}},$$
$$\widetilde{\theta} = [\widetilde{\theta}_1, \widetilde{\theta}_2, \ldots, \widetilde{\theta}_n]^{\mathsf{T}}, \xi = [\xi_1, \xi_2, \ldots, \xi_n]^{\mathsf{T}}, \widehat{\eta} = [\widehat{\eta}_1, \widehat{\eta}_2, \ldots, \widehat{\eta}_n]^{\mathsf{T}}.$$

Then, (3.32) can be rewritten in a compact form, as

$$\begin{aligned} \widetilde{\rho} &= \mathbf{1}_n \otimes r - \rho, \\ \widetilde{\theta} &= \mathbf{1}_n \otimes \frac{2\pi}{n} - (-L\theta + \xi), \end{aligned} \tag{3.33}$$

with

$$L = \begin{bmatrix} 1 & -1 & 0 & \ldots & 0 \\ 0 & 1 & -1 & 0 & \ldots \\ 0 & 0 & \ldots & \ldots & \ldots \\ -1 & 0 & \ldots & \ldots & 1 \end{bmatrix}. \tag{3.34}$$

Its derivative is

$$\begin{cases} \dot{\widetilde{\rho}} = -k_2\widetilde{\rho} + \widetilde{\eta}, \\ \dot{\widetilde{\theta}} = -L\widetilde{\theta} + \widetilde{\omega}, \end{cases} \tag{3.35}$$

with $\widetilde{\eta} = -\widehat{\eta}$, $\widetilde{\omega} = -L\omega$. Condition $\lim_{t\to\infty} \widehat{\eta}_i(t) = 0$ together with $\lim_{t\to\infty} \widehat{\omega}_i(t) = 0$ implies $\lim_{t\to\infty} \widetilde{\eta}_i(t) = 0$, $\lim_{t\to\infty} \widetilde{\omega}_i(t) = 0$.

It follows from (3.10) that $\mathbf{1}_n^{\mathsf{T}}\xi = 2\pi$. Moreover, from (3.32), the equality holds, giving

$$\mathbf{1}_n^{\mathsf{T}}\widetilde{\theta} = \mathbf{1}_n^{\mathsf{T}}\left(\mathbf{1}_n \otimes \frac{2\pi}{n} - (-L\theta + \xi)\right) = 2\pi - 2\pi = 0. \tag{3.36}$$

According to [34], one has

$$\min_{\mathbf{1}_n^{\mathsf{T}}\widetilde{\theta}=0, \widetilde{\theta}\neq 0} \frac{\widetilde{\theta}^{\mathsf{T}} L \widetilde{\theta}}{\|\widetilde{\theta}\|^2} = \lambda_2(L) > 0, \tag{3.37}$$

with $\lambda_2(L)$ being the second smallest eigenvalue of L. Let $\chi = \min\{k_2, \lambda_2(L)\} > 0$, and define

$$\boldsymbol{\varepsilon} := \begin{bmatrix} \widetilde{\rho} \\ \widetilde{\theta} \end{bmatrix}, \widetilde{\gamma} := \begin{bmatrix} \widetilde{\eta} \\ \widetilde{\omega} \end{bmatrix}. \tag{3.38}$$

Clearly, one has $\lim_{t\to\infty} \widetilde{\gamma}(t) = 0$ if $\lim_{t\to\infty} \widetilde{\eta}(t) = 0$ and $\lim_{t\to\infty} \widetilde{\omega}(t) = 0$. Pick a Lyapunov function $V(\boldsymbol{\varepsilon}) = \boldsymbol{\varepsilon}^{\mathsf{T}}\boldsymbol{\varepsilon}/2$, satisfying

$$\begin{aligned}\dot{V}(\boldsymbol{\varepsilon}) &= \widetilde{\rho}^{\mathrm{T}}\dot{\widetilde{\rho}} + \widetilde{\theta}^{\mathrm{T}}\dot{\widetilde{\theta}} \\ &= -k_2\widetilde{\rho}^{\mathrm{T}}\widetilde{\rho} + \widetilde{\rho}^{\mathrm{T}}\widetilde{\eta} + \widetilde{\theta}^{\mathrm{T}}\big(-L\widetilde{\theta} + \widetilde{\omega}\big) \\ &\leq -k_2\left\|\widetilde{\rho}\right\|^2 - \lambda_2(L)\left\|\widetilde{\theta}\right\|^2 + \boldsymbol{\varepsilon}^{\mathrm{T}}\widetilde{\gamma} \\ &\leq -\chi\big(\left\|\widetilde{\rho}\right\|^2 + \left\|\widetilde{\theta}\right\|^2\big) + \left\|\boldsymbol{\varepsilon}\right\| \cdot \left\|\widetilde{\gamma}\right\| \\ &\leq -\chi(1-c_2)\left\|\boldsymbol{\varepsilon}\right\|^2 + \left\|\boldsymbol{\varepsilon}\right\|\big(\left\|\widetilde{\gamma}\right\| - \chi c_2\left\|\boldsymbol{\varepsilon}\right\|\big)\end{aligned} \tag{3.39}$$

with $c_2 \in (0, 1)$. Then, one has that $\dot{V}(\boldsymbol{\varepsilon}) \leq -\chi(1-c_2)\left\|\boldsymbol{\varepsilon}\right\|^2,\ \forall \left\|\boldsymbol{\varepsilon}\right\| \geq \frac{\|\widetilde{\gamma}\|}{\chi c_2}$.

Analogously, the closed-loop system of $\boldsymbol{\varepsilon}$ is ISS with respect to $\widetilde{\gamma}$ and satisfies

$$\|\boldsymbol{\varepsilon}(t)\| \leq \beta(\|\boldsymbol{\varepsilon}(0\|, t)) + \gamma\big(\sup_{0\leq\tau\leq t}\|\widetilde{\gamma}(\tau)\|\big),\ \forall t > 0, \tag{3.40}$$

where $\beta(\cdot)$ is a $\mathcal{KL}$ function and $\kappa_2 = 1/\chi c_2$ is the gain of the system (see [33]). It follows from $\mathcal{KL}$ function $\beta(\|\boldsymbol{\varepsilon}(0\|, t))$ and (3.38) that

$$\lim_{t\to\infty}\beta(\|\boldsymbol{\varepsilon}(0\|, t)) = 0,\ \lim_{t\to\infty}\|\widetilde{\gamma}(t)\| = 0. \tag{3.41}$$

Accordingly, one has $\lim_{t\to\infty}\boldsymbol{\varepsilon}(t) = 0$, i.e.,

$$\lim_{t\to\infty}\widetilde{\rho}_i(t) = 0,\quad \lim_{t\to\infty}\widetilde{\theta}_i(t) = 0, \tag{3.42}$$

which implies that $\lim_{t\to\infty}\rho_i = r$ and $\lim_{t\to\infty}(\theta_{i+} - \theta_i + \xi_i) = \frac{2\pi}{n}$, i.e.,

$$\lim_{t\to\infty}\theta_{i+} - \theta_i + \xi_i = \lim_{t\to\infty}\theta_{j+} - \theta_j + \xi_j,\ i, j \in \mathcal{V}, i \neq j, \tag{3.43}$$

subject to $\lim_{t\to\infty}\widehat{\eta}_i(t) = 0$ and $\lim_{t\to\infty}\widehat{\omega}_i(t) = 0$. This completes the proof. □

Remark 3.2 Note that the conditions $\lim_{t\to\infty}\widehat{\eta}_i(t) = 0$ and $\lim_{t\to\infty}\widehat{\omega}_i(t) = 0$, i.e., $\lim_{t\to\infty}\widetilde{e}_i(t) = 0$, ensure the decoupling of high-level collective behavior control from low-level underactuated dynamics control, which are assumed to be satisfied for Problems 3.1 and 3.2. Consequently, Problem 3.3 can be addressed using the previously proposed controller from [30].

Remark 3.3 Theorem 3.1 indicates that vessels switch to Stage 2 if Problem 3.2 is resolved, i.e., $\lim_{t\to\infty}z_i = 0$. To speed up this process in real-world experiments, the switching condition between Stages 1 and 2 should be triggered when $|z_i| \leq \sigma$ for $i \in \mathcal{V}$, where $\sigma > 0$ is a predefined threshold that ensures the asymptotic convergence of $|z_i|$ in (3.19).

3.5 Algorithm Verification

In this section, the proposed fencing control scheme (3.13), (3.17), and (3.25) has been implemented on our lake-based multi-USV platform, which consists of three HUSTER-12 vessels: a base control station (BCS), a virtual moving target vessel, and a USV dock.

As depicted in Fig. 3.3, each HUSTER-12 USV is 1.6 m in length, with dynamic model parameters in (3.3) defined as follows: $m_{11} = 1000.00$ kg, $m_{22} = 0.60$ kg, $m_{33} = 263.16\ \text{kg} \cdot \text{m}^2$, $d_{11} = 980.40\ \text{kg} \cdot \text{s}^{-1}$, $d_{22} = 0.59\ \text{kg} \cdot \text{s}^{-1}$, $d_{33} = 257.61\ \text{kg} \cdot \text{m}^2 \cdot \text{s}^{-1}$. Each vessel is equipped with an onboard differential GPS (DGPS) receiver (NovAtel-OEM615), a DGPS wireless sensor (E32-400T20S), two DGPS antennas, a 433 M wireless sensor (E32-433T30S), an accelerometer–gyroscope (AG) chip (MPU6050), an embedded controller (STM32F407VGT6), a motor driver (SEAKING-V3), a high-power battery (RUIYI-48), and a waterjet propeller (HJ-064). The AG chip is integrated for vessel navigation. Typically, using the onboard DGPS receiver and DGPS station, each USV determines its heading angle and position with an accuracy of ± 2 cm, which is sent to the embedded controller to generate actuator signals. Through the communication of intermittent neighboring information via the 433M wireless sensor and receiver, each USV calculates its control signals, which are executed by the motor driver. This driver regulates the waterjet engine (2 kW) and rudder (with a range of $\pm 15°$) to accomplish the two-stage fencing control mission. The status trajectories of the vessels and the corresponding controller

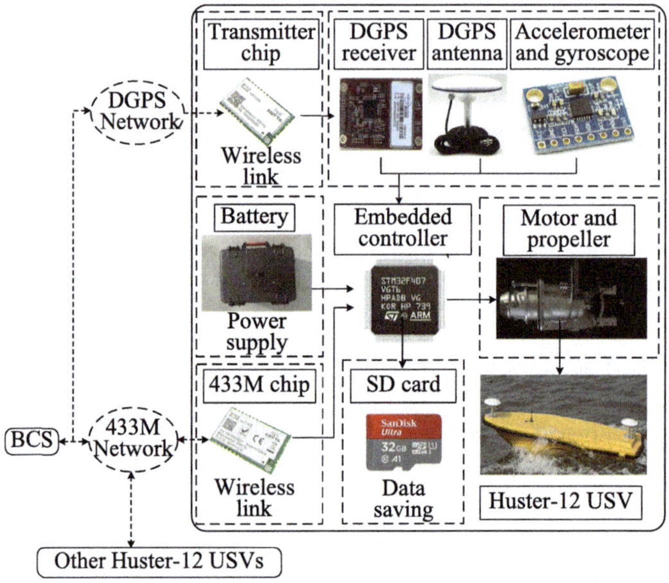

Fig. 3.3 Hardware structure of the multi HUSTER-12s USVs

Table 3.1 Main components in the multi-USV system

Component/hardware	Parameter
NovAtel-OEM615	Accuracy: ±2 cm
E32-400T20S	Power: 10–20 dBm, distance: 3 km
E32-433T30S	Power: 21–30 dBm, distance: 8 km
MPU6050	Gyroscope range: ±250, 500, 1000, 2000°/s, acceleration range: ±2 g, ±4 g, ±8 g, ±16 g
STM32F407VGT6	Flash: 1024 kB, frequency: 72 M, pin:100
SEAKING-V3	Voltage: 5–12 s Lipo (21–50.4 V), current: 130 A
RUIYI-48	Capacity: 48 V, 30 Ah
HJ064	Power: 3.6 kW, thrust: 115 N

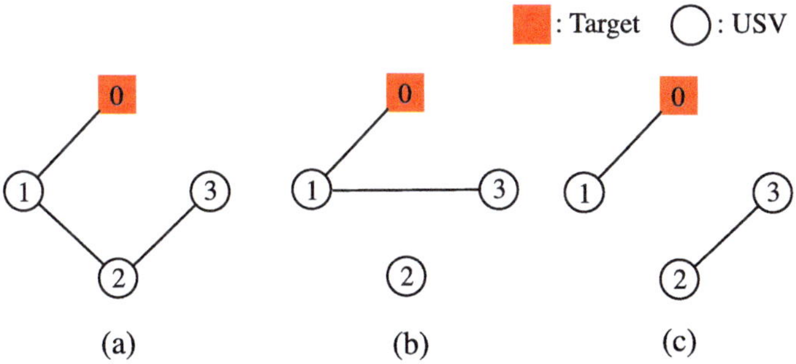

Fig. 3.4 Switching inter-vessel communication topologies of the multi-USV system. Circles 1, 2, and 3 are the vessels, and red rectangle 0 is the virtual motional target

parameters are logged by an onboard SD card (32 GB). Table 3.1 provides a summary of the main components and hardware specifications for the multi-USV system.

In this experiment, the three vessels start from random positions within a $[20 \times 20]$ m^2 lake area. The sensing radius R is set to 18 m, and the encircling radius r is also set to 18 m. Initially, the vessels have states of $u_i = 0$ m/s, $v_i = 0$ m/s, and $r_i = 0$ rad/s. The control signals are calculated and updated by the embedded controller at a frequency of 4 Hz. For the moving target $q_0(t)$, the initial position is set to $q_0 = [-9510, 12010]^{\top}$ cm, and the velocity is $v_0 = [-50, 50]^{\top}$ cm/s, both defined in the East–North coordinate system. The intermittent communication network topology among the USVs is illustrated in Fig. 3.4, which periodically switches between configurations (a), (b), and (c) with a switching rate of $\mu = 0.9$. When the distance between the visual target and any follower USV exceeds r, the follower loses direct access to the virtual moving target. In such cases, the follower USVs exchange their estimated positions $\widehat{q}_0^i$ with neighboring vessels via the onboard 433 M wireless communication network if the target is outside their sensing range. The controller parameters for (3.13), (3.17), and (3.25) are set to $k_1 = 0.1$, $k_2 = 2$, $c_2 = 0.2$, $c_3 = 0.2$, and $\gamma = 0.5$, ensuring they meet the conditions given in (3.15).

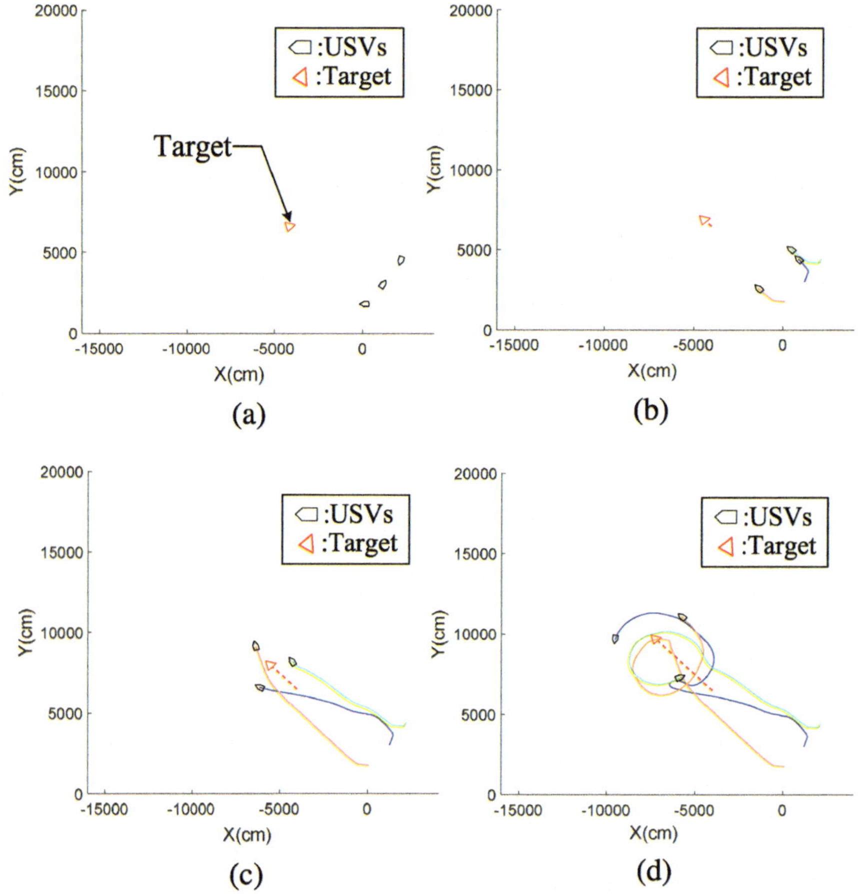

Fig. 3.5 Two-stage fencing process of a virtual moving target with the multi-USV system. The colored lines are the moving trajectories of the three vessels. **a** Initial position of the USVs; **b** USVs begin to chase the virtual motional target; **c** The target has been fenced in Stage 1; **d** USVs evenly rotate around the moving target in Stage 2

The trajectories of the vessels during the two-stage fencing control process while following the moving target $q_0(t)$ are shown in Fig. 3.5. As seen in Fig. 3.6, the distance between each vessel and the target stabilizes around 12 m. Additionally, the relative angle between each pair of neighboring vessels uniformly converges to 120° within 115 s, indicating that the three vessels effectively capture and rotate symmetrically around the moving target at the center within the intermittent communication network. Regarding estimation errors, the evolution of the estimated state errors $\widetilde{q}_0^i$ and $\widetilde{v}_0^i$ for each vessel ($i = 1, 2, 3$), as determined by the observer in (3.13), is plotted in Fig. 3.7. The estimation errors are observed to settle to zero within 20 s, which validates the effectiveness of the designed observer in (3.13).

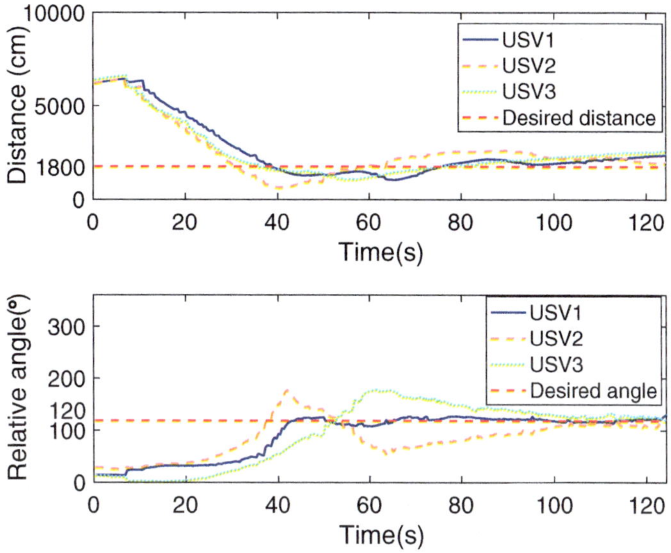

Fig. 3.6 Evolution of the distance and the relative angle from the vessels to the virtual moving target, respectively

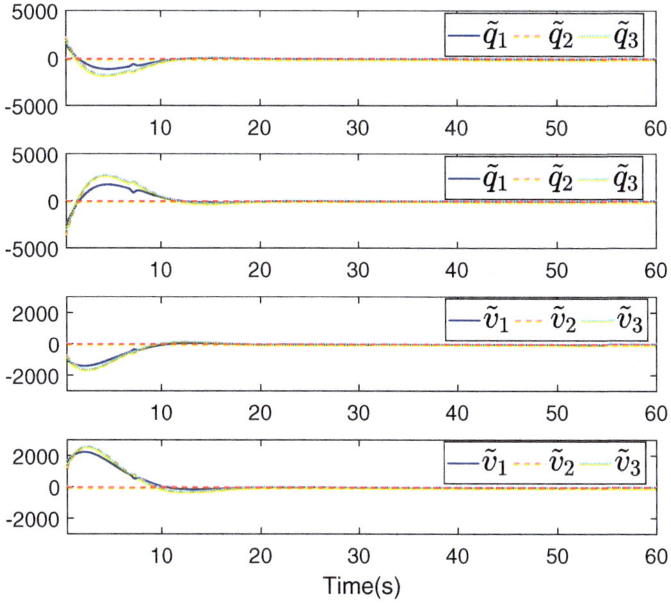

Fig. 3.7 Evolution of the estimated position errors $\widetilde{q}_i(t)$ and velocity errors $\widetilde{v}_i(t)$ toward the a virtual moving target, respectively

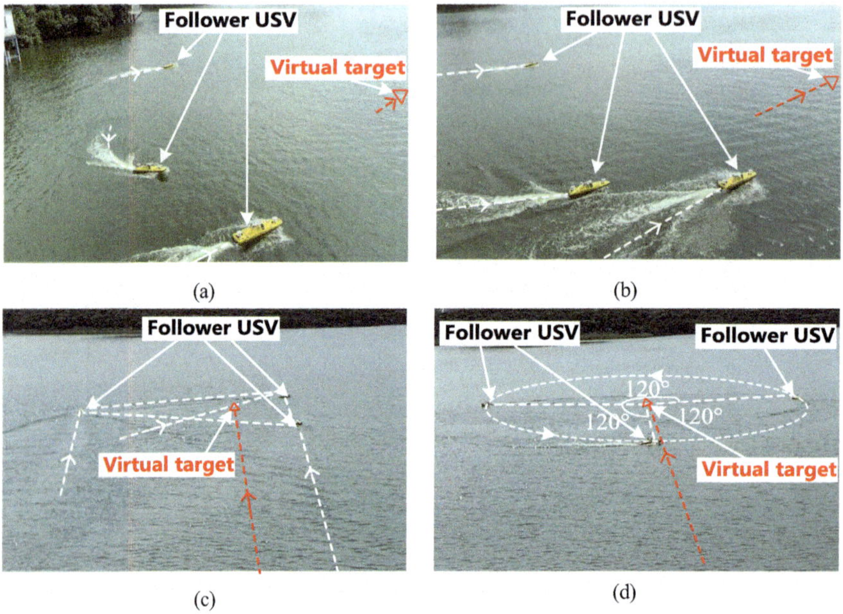

Fig. 3.8 A real snapshot of the two-stage equally surrounding control to follow the virtual moving target

To provide a visualization of the fencing control performance, a real snapshot of the vessels in an equally cyclic fencing formation is shown in Fig. 3.8. These experiments demonstrate the effectiveness of the proposed two-stage control laws (3.17) and (3.25), as well as the feasibility of Theorems 3.1 and 3.2. Notably, since the experiments were conducted with intermittent communication (as depicted in Fig. 3.4), which is not addressed by existing studies such as [8–15], the proposed control method's superiority is further validated in achieving the target encircling mission with evenly rotating phases, unknown target velocities, and dynamic communication topologies.

It is important to note that it typically takes a longer time for the USVs to achieve an even encirclement of a moving target, and the fencing trajectories exhibit more oscillations. This is mainly due to the increased external disturbances from water currents and a higher communication packet loss rate caused by the need to continuously fence the moving target. Despite these challenges, the overall performance remains satisfactory.

3.6 Conclusion

In this chapter, a distributed fencing controller via input-to-state stability is proposed to address a two-stage equal fencing problem of multi-USV systems. The key advantage of the proposed controller is its ability to handle joint dwelling connectivity in the inter-agent communication graph, making it more practical and reducing communication costs between vessels. Importantly, the chapter establishes conditions that ensure the asymptotic stability of the closed-loop multi-USV system under this control strategy. The effectiveness of the proposed controller is validated through real-world lake experiments involving three HUSTER-12 USVs, demonstrating satisfactory performance. This work bridges the gap between theoretical design and practical implementation of multi-USV systems by addressing challenges faced in real-world scenarios. Future research will focus on extending the fencing control strategy to handle multiple moving targets with significant velocity variations.

References

1. M. Bibuli, G. Bruzzone, M. Caccia, L. Lapierre, Path-following algorithms and experiments for an unmanned surface vehicle. J. Field Robot. **26**(8), 669–688 (2009)
2. M. Caccia, G. Bruzzone, R. Bono, A practical approach to modeling and identification of small autonomous surface craft. IEEE J. Ocean. Eng. **33**(2), 133–145 (2008)
3. W.B. Klinger, I.R. Bertaska, K.D. von Ellenrieder, M.R. Dhanak, Control of an unmanned surface vehicle with uncertain displacement and drag. IEEE J. Ocean. Eng. **42**(2), 458–476 (2017)
4. E. Lefeber, K.Y. Pettersen, H. Nijmeijer, Tracking control of an underactuated ship. IEEE Trans. Control Syst. Technol. **11**(1), 52–61 (2003)
5. H. Ashrafiuon, K.R. Muske, L.C. McNinch, R.A. Soltan, Sliding-mode tracking control of surface vessels. IEEE Trans. Ind. Electron. **55**(11), 4004–4012 (2008)
6. W. He, Z. Yin, C. Sun, Adaptive neural network control of a marine vessel with constraints using the asymmetric barrier Lyapunov function. IEEE Trans. Cybern. **47**(7), 1641–1651 (2017)
7. Y.L. Wang, Q.L. Han, Network-based fault detection filter and controller coordinated design for unmanned surface vehicles in network environments. IEEE Trans. Ind. Inform. **12**(5), 1753–1765 (2016)
8. Z. Peng, D. Wang, Z. Chen, X. Hu, W. Lan, Adaptive Dynamic Surface Control for Formations of Autonomous Surface Vehicles With Uncertain Dynamics. IEEE Trans. Control Syst. Technol. **21**(2), 513–520 (2013)
9. B. Xiao, X. Yang, X. Huo, A novel disturbance estimation scheme for formation control of ocean surface vessels. IEEE Trans. Ind. Electron. **64**(6), 4994–5003 (2017)
10. Z. Peng, D. Wang, T. Li, M. Han, Output-feedback cooperative formation maneuvering of autonomous surface vehicles with connectivity preservation and collision avoidance. IEEE Trans. Cybern. **50**(6), 2527–2535 (2019)
11. J. Ghommam, F. Mnif, Coordinated path-following control for a group of underactuated surface vessels. IEEE Trans. Ind. Electron. **56**(10), 3951–3963 (2009)
12. K.D. Do, Synchronization motion tracking control of multiple underactuated ships with collision avoidance. IEEE Trans. Ind. Electron. **63**(5), 2976–2989 (2016)
13. T. Li, R. Zhao, C.L.P. Chen, L. Fang, C. Liu, Finite-time formation control of under-actuated ships using nonlinear sliding mode control. IEEE Trans. Cybern. **48**(11), 3243–3253 (2018)

14. Z. Peng, J. Wang, D. Wang, Distributed containment maneuvering of multiple marine vessels via neurodynamics-based output feedback. IEEE Trans. Ind. Electron. **64**(5), 3831–3839 (2017)
15. Z. Peng, J. Wang, D. Wang, Containment maneuvering of marine surface vehicles with multiple parameterized paths via spatial-temporal decoupling. IEEE/ASME Trans. Mechatron. **22**(2), 1026–1036 (2017)
16. H. Yamaguchi, A cooperative hunting behavior by mobile-robot troops. Int. J. Robot. Res. **18**(9), 931–940 (1999)
17. J. Ni, S.X. Yang, Bioinspired neural network for real-time cooperative hunting by multirobots in unknown environments. IEEE Trans. Neural Netw. **22**(12), 2062–2077 (2011)
18. R. Sepulchre, D.A. Paley, N.E. Leonard, Stabilization of planar collective motion: All-to-all communication. IEEE Trans. Autom. Control **52**(5), 811–824 (2007)
19. J.A. Marshall, M.E. Broucke, B.A. Francis, Formations of vehicles in cyclic pursuit. IEEE Trans. Autom. Control **49**(11), 1963–1974 (2004)
20. A. Sinha, D. Ghose, Generalization of nonlinear cyclic pursuit. Automatica **43**(11), 1954–1960 (2007)
21. T.-H. Kim, T. Sugie, Cooperative control for target-capturing task based on a cyclic pursuit strategy. Automatica **43**(8), 1426–1431 (2007)
22. N. Ceccarelli, M. Di Marco, A. Garulli, A. Giannitrapani, Collective circular motion of multi-vehicle systems. Automatica **44**(12), 3025–3035 (2008)
23. D. Li, G. Ma, W. He, S.S. Ge, T.H. Lee, Cooperative circumnavigation control of networked microsatellites. IEEE Trans. Cybern. **50**(10), 4550–4555 (2019)
24. Y. Lan, G. Yan, Z. Lin, Distributed control of cooperative target enclosing based on reachability and invariance analysis. Syst. Control Lett. **59**(7), 381–389 (2015)
25. X. Yu, L. Liu, Distributed circular formation control of ring-networked nonholonomic vehicles. Automatica **68**, 92–99 (2016)
26. R. Zheng, Y. Liu, D. Sun, Enclosing a target by nonholonomic mobile robots with bearing-only measurements. Automatica **53**, 400–407 (2015)
27. R. Li, Y. Shi, Y. Song, Localization and circumnavigation of multiple agents along an unknown target based on bearing-only measurement: A three dimensional solution. Automatica **94**, 18–25 (2018)
28. X. Yu, L. Liu, Cooperative control for moving-target circular formation of nonholonomic vehicles. IEEE Trans. Autom. Control **62**(7), 3448–3454 (2017)
29. X. Yu, X. Xu, L. Liu, G. Feng, Circular formation of networked dynamic unicycles by a distributed dynamic control law. Automatica **89**, 1–7 (2018)
30. B. Liu, Z. Chen, H.-T. Zhang, X. Wang, T. Geng, H. Su, J. Zhao, Collective dynamics and control for multiple unmanned surface vessels. IEEE Trans. Control Syst. Technol. **28**(6), 2540–2547 (2019)
31. Y. Hong, J. Hu, L. Gao, Tracking control for multi-agent consensus with an active leader and variable topology. Automatica **42**(7), 1177–1182 (2006)
32. K.H. Khalil, *Nonlinear Systems* (Prentice Hall, Upper Saddle River, 2002)
33. A. Isidori, *Nonlinear control systems II* (Springer Science & Business Media, Berlin, Germany, 2012)
34. R. Olfati-Saber, R.M. Murray, Consensus problems in networks of agents with switching topology and time-delays. IEEE Trans. Autom. Control **49**(9), 1520–1533 (2004)

Chapter 4
Distributed Fencing Control via Output Regulation Theory

4.1 Introduction

In Chaps. 2 and 3, we have explored the themes of "basic fencing" and "complex fencing" with an evenly rotating formation based on the neighboring angle repulsion and input-to-state stability, respectively. Although Chaps. 2 and 3 addressed a moving target fencing and formation evolution, they only are applied to a constant-speed target. However, in practical applications, the varying-velocity target is more common, which has a more practical value in extremely dangerous situations. Therefore, this chapter continues to focus on the theme of "complex fencing". Following the direction of the previous two chapters, we further propose a distributed controller via output regulation theory, which aims at solving a more general problem of cooperative fencing of targets with a time-varying velocity. In what follows, we will introduce the related background.

Inspired by the highly coordinated behaviors of bird flocks, fish schools, insect colonies, and mammal herds in nature, recent research into the cooperative control of MRSs has gained momentum. Such natural coordination offers key advantages like efficiency, broad coverage, and cost-effectiveness, making it applicable to industries, engineering, and social network systems. The emergence of strong cooperative behaviors has long been a focal point in MRS control research. Noteworthy advancements have been made in recent years, particularly in formation control [1, 2], synchronization/consensus methods [3, 4], containment strategies [5, 6], and circular motion control [7, 8].

In the realm of collective circular motion control, the issue of target fencing has garnered growing interest. Early research on the target-fencing problem in multi-robot systems (MRS) primarily concentrated on confining a target within the trajectories of well-informed robots [9–11] that had direct access to target information. However, this scenario is not common in practical applications. To address this, a more realistic cooperative strategy was proposed in [12], where only a subset of the robots has access to the target.

B.-B. Hu and H.-T. Zhang, *Multi-Robot Systems*, Cognitive Intelligence and Robotics,
https://doi.org/10.1007/978-981-97-9214-6_4

While previous works like [9–12] have achieved target encirclement using moving trajectories, they do not consistently ensure that the target remains fully enclosed at every instant. This observation inspires another research avenue known as the target-fencing problem, where the goal is to continuously enclose the target within the convex hull of the entire group of robots. In early studies on fencing control for multi-robot systems (MRS), all robots had fixed labels, with predetermined neighbors and relative positions. This problem has been addressed for first-order MRSs [13] and second-order MRSs [14]. Later, this concept was expanded to multi-target fencing scenarios for second-order nonlinear MRSs [15] and even multiple unmanned surface vessels [16].

However, label-fixed fencing controllers often struggle with increasingly complex tasks in unpredictable environments. As missions grow more intricate, these controllers must constantly recalculate the fixed relative positions among robots, which adds to the computational burden and reduces overall efficiency. In unpredictable settings, these controllers may also take longer and consume more energy to achieve effective cooperation. For example, if a robot starts in a position that is significantly different from its designated location, it must navigate through the formation to reach its assigned spot, which can be inefficient. Moreover, if a robot fails during the process, the predefined positions make it impossible to maintain the fencing formation. This highlights the need for more adaptable label-free fencing controllers, where robots do not have predefined labels [17]. Various solutions have been proposed to address this challenge: a limit-cycle-based approach for maintaining a stationary target [18], a hierarchical structure with an output regulation method for surrounding a stationary target [19], and a cooperative controller using attractive, repulsive, and rotational forces to enclose a specified stationary target [20]. Additionally, research has extended to fencing a moving target with constant velocity for first-order MRSs [21], second-order MRSs [22, 23], and multiple unmanned surface vessels [24].

To date, most research [18–20] on label-free fencing has focused on first-order MRSs and has not thoroughly addressed the evolution of the formation throughout the entire fencing process. This consideration is crucial for practical applications such as unmanned system convoy protection, reconnaissance, and patrol. Although some recent studies [21–24] have explored rigid formations for constant-velocity targets, the more complex problem of fencing a moving target with varying velocity remains unresolved. Therefore, it is both urgent and challenging to develop a label-free controller for second-order MRSs that ensures collision-free rigid-formation fencing for targets with variable velocities. The main contribution of this chapter is two-fold.

1. Design a label-free controller for a second-order MRS to collaboratively fence a complex target of changing velocity within their convex hull, all while maintaining a fixed formation.
2. Ensure both collision avoidance between robots and convergence for a nonlinear MRS that is subject to complex dynamics, strong inter-robot couplings, and time-varying network topologies.

The primary challenge addressed in this chapter is the strong nonlinear couplings introduced by the label-free nature of the system combined with a varying-velocity target, which makes two significant contributions. (i) It extends beyond previous studies that focused on label-free fencing for static targets [18–20] and constant-velocity targets [21–24] by treating the target as an exosystem. This approach leads to the development of a novel label-free controller that incorporates a dynamic regulator and inter-robot repulsion to tackle the more complex problem of fencing a target with variable velocity. (ii) Second, by integrating inter-robot repulsion into the internal model, this chapter ensures both rigid formation and collision avoidance. Notably, the method can also address the constant-velocity target fencing problem as a special case by using the dynamic regulator within the internal model.

The organization of this chapter is as follows. Section 4.2 outlines the preliminaries and defines the primary problem tackled in this chapter. Section 4.3 formulates the target-fencing control law and presents the principal technical results. Section 4.4 conducts numerical simulations to demonstrate the effectiveness of the proposed approach. Section 4.5 provides the concluding remarks.

4.2 Problem Formulation of Fencing a Varying-Velocity Target

In this section, we first consider an MRS comprising n robots, where each robot operates under second-order dynamics in Cartesian coordinates,

$$\begin{aligned} \dot{x}_i(t) &= v_i(t) \\ \dot{v}_i(t) &= u_i(t), \end{aligned} \tag{4.1}$$

where $x_i(t)$ and $v_i(t)$ indicate the position and velocity of robot i, respectively, while $u_i(t)$ represents the control input. The MRS topology is represented by $\mathcal{G} = (\mathcal{V}, \mathcal{E})$, where $\mathcal{V} = \{1, 2, \ldots, n\}$ denote the set of nodes and $\mathcal{E} \subseteq \mathcal{V} \times \mathcal{V}$ denotes the set of edges. $\mathcal{N}_i$ is defined as the sensing neighborhood set of robot i in $\mathcal{V}$, i.e.,

$$\mathcal{N}_i(t) := \{k \in \mathcal{V}, k \neq i \mid \|x_{i,k}(t)\| \leq R\} \tag{4.2}$$

with a detectable range $R \in (r, \infty)$ and a specified safe distance $r \in \mathbb{R}^+$. The relative position between robots i and k is given by $x_{i,k}(t) := x_i(t) - x_k(t)$, where $i \neq k \in \mathcal{V}$. From the condition that the relative distances $|x_{i,k}|$ are both limited and variable, one has that the neighborhood set $\mathcal{N}_i$ for robot i changes over time. As a result, the topology $\mathcal{G}$ may also vary dynamically. This situation is more complex than cases with fixed topologies and predetermined robot labels (see, e.g., [13–16]).

Then, we consider a moving target satisfying

$$\begin{bmatrix} \dot{x}_d(t) \\ \dot{v}_d(t) \end{bmatrix} = S \begin{bmatrix} x_d(t) \\ v_d(t) \end{bmatrix} \tag{4.3}$$

with the dynamic matrix

$$S = \begin{bmatrix} 0 & 1 \\ s_1 & 0 \end{bmatrix} \otimes I_2 \tag{4.4}$$

where $s_1 \leq 0$ is a constant, and $x_d \in \mathbb{R}^2$, $v_d \in \mathbb{R}^2$ represent the position and velocity of the target, respectively. When $s_1 = 0$ in (4.4), the target in (4.3) moves with a constant velocity, as discussed in [22]. For $s_1 < 0$, the target moves periodically with a varying velocity.

Let co(x) be the convex hull of all the robots $x = [x_1, \ldots, x_n]^{\mathrm{T}}$, i.e.,

$$\mathrm{co}(x) := \left\{ \sum_{i \in \mathcal{V}} \lambda_i x_i : \lambda_i \geq 0, \forall i \text{ and } \sum_{i \in \mathcal{V}} \lambda_i = 1 \right\}, \tag{4.5}$$

which implies that the distance from the target to the convex hull co(x) is given by $P_{x_d}(x) := \min_{s \in \mathrm{co}(x)} \|x_d - s\|$ (see, e.g., [20]). Accordingly, we can give the following definition.

Definition 4.1 (*Label-free rigid-formation fencing* [20]) An MRS $\mathcal{V}$ with dynamic (4.1) asymptotically fences a target governed by the dynamic (4.3) into a convex hull (4.5) formed by the arbitrarily labeled robots, while ensuring a collision-free rigid formation, if the following three claims are met:

$$\begin{aligned} &(1) \lim_{t \to \infty} P_{x_d(t)}(x(t)) = 0, \\ &(2) \lim_{t \to \infty} v_i(t) - v_d(t) = 0, \forall i \in \mathcal{V}, \\ &(3) \|x_{i,k}(t)\| > r, \forall t \geq 0, \ \forall i \neq k \in \mathcal{V}. \end{aligned} \tag{4.6}$$

In Definition 4.1, Claim (1) asserts that the target x_d is enclosed within the convex hull co(x) formed by arbitrarily labeled robots x_i, $i \in \mathcal{V}$. From Claim (2), where $\lim_{t \to \infty}(v_i(t) - v_d(t)) = 0$, it follows that $\lim_{t \to \infty}(x_i(t) - x_d(t)) = d_i$, with a constant vector $d_i \in \mathbb{R}^2$. This indicates that the relative positions between the robots and the target remain consistent, thereby ensuring that the MRS maintains a fixed formation. Finally, Claim (3) guarantees that the distance $|x_{i,k}(t)|$ between any pair of robots always exceeds r, ensuring collision avoidance among the robots. Combining Claims (1) and (3) together yields the fencing formation. Claim (2) introduces an additional condition needed to maintain a rigid formation during fencing. Furthermore, note that Claims (1)–(3) do not explicitly predefine fixed labels or inter-robot relative positions for specific robots, indicating that the labeling of robots in the fencing formation is flexible, resulting in label-free fencing.

To illustrate the main problem addressed in this chapter more clearly, an example of label-free fencing is presented in Fig. 4.1. In this figure, all patterns 1–3 demonstrate successful fencing of a moving target, highlighting that no specific labels are

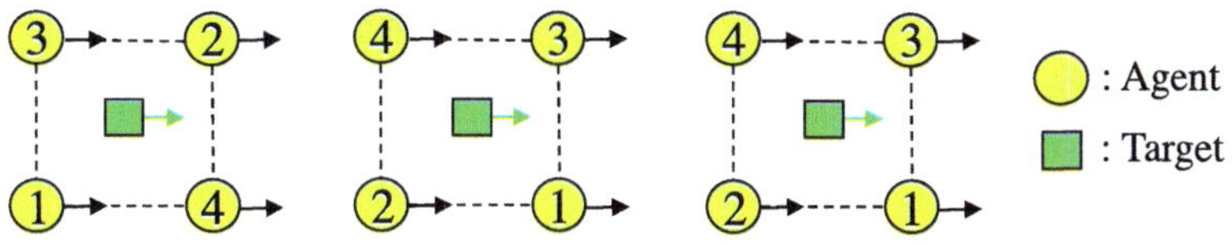

(a) Pattern 1 (b) Pattern 2 (c) Pattern 3

Fig. 4.1 Example of label-free fencing for a motional target

needed for individual robots. Now, the main problem addressed in this chapter can be formally introduced as follows.

Problem 4.1 *The objective is to design a cooperative label-free controller*

$$u_i := f(x_i, v_i, x_d, x_k), i \in \mathcal{V}, k \in \mathcal{N}_i, \tag{4.7}$$

for an MRS (4.1) *and a target of variational velocity in* (4.3) *to achieve label-free rigid-formation fencing, as shown in Definition 4.1.*

4.3 Internal-Model Controller Design and Convergence Analysis

Let the relative position between an arbitrary robot $i \in \mathcal{V}$ and the target be given by

$$e_i(t) := x_i(t) - x_d(t). \tag{4.8}$$

Now, by omitting (t) for simplicity, a cooperative fencing controller with dynamic feedback regulation for $i \in \mathcal{V}$ is proposed as follows:

$$\begin{aligned} u_i &= -\left(\begin{bmatrix} k_1 & k_2 \end{bmatrix} \otimes I_2\right)\begin{bmatrix} x_i \\ v_i \end{bmatrix} - \left(\begin{bmatrix} k_3 & k_4 \end{bmatrix} \otimes I_2\right)\begin{bmatrix} \epsilon_i \\ \zeta_i \end{bmatrix} + k_5 \sum_{k \in \mathcal{N}_i} \alpha(\|x_{i,k}\|)\frac{x_{i,k}}{\|x_{i,k}\|}, \\ \begin{bmatrix} \dot{\epsilon}_i \\ \dot{\zeta}_i \end{bmatrix} &= \left(\begin{bmatrix} 0 & 1 \\ s_1 & 0 \end{bmatrix} \otimes I_2\right)\begin{bmatrix} \epsilon_i \\ \zeta_i \end{bmatrix} + \left(\begin{bmatrix} 0 \\ 1 \end{bmatrix} \otimes I_2\right)\left(e_i - k_5 \sum_{k \in \mathcal{N}_i} \alpha(\|x_{i,k}\|)\frac{x_{i,k}}{\|x_{i,k}\|}\right), \end{aligned} \tag{4.9}$$

where $k_1, k_2, k_3, k_4, k_5 \in \mathbb{R}^+$ are positive parameters, and $\epsilon_i, \zeta_i \in \mathbb{R}^2$ represent the states of the internal model in (4.9) for robot i. The neighborhood set $\mathcal{N}i$ is defined in (4.2). The term $\sum_{k \in \mathcal{N}_i}\{\alpha(\|x_{i,k}\|)x_{i,k}/\|x_{i,k}\|\}$ represents the inter-robot repulsion, where the continuous function $\alpha(\cdot)$ (see, e.g., [20]) satisfies

$$\alpha(s) = 0, \forall s \in [R, \infty), \lim_{s \to r^+} \alpha(s) = \infty. \tag{4.10}$$

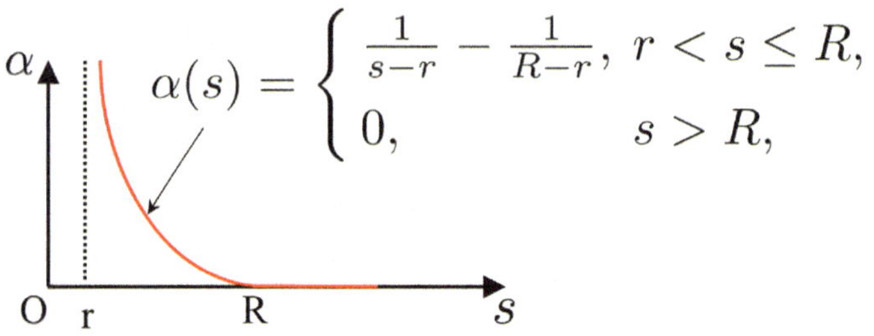

Fig. 4.2 Illustration of the potential function $\alpha(s)$ in (4.12)

Here, the detectable range $R \in (r, \infty)$ and the specified safe distance r are given in (4.2). In the following discussions, for conciseness of presentation, we denote

$$\eta_i := \sum_{k \in \mathcal{N}_i} \{\alpha(\|x_{i,k}\|) \frac{x_{i,k}}{\|x_{i,k}\|}\}. \tag{4.11}$$

Remark 4.1 Note that the function $\alpha(s)$ in (4.10) is monotonically decreasing for $s \in (r, R]$ and equals 0 for $s \in (R, \infty)$. This indicates that $\alpha(s)$ is Lipschitz continuous over (r, ∞). An example of $\alpha(\cdot)$ that satisfies (4.10) is illustrated in Fig. 4.2:

$$\alpha(s) = \begin{cases} \frac{1}{s-r} - \frac{1}{R-r}, & r < s \leq R, \\ 0, & s > R. \end{cases} \tag{4.12}$$

Remark 4.2 As shown in the controller (4.9), the position state x_d, although known to each robot, does not represent the final convergent steady state, which is different from some centralized controllers (such as those using a star topology). Instead, the steady states of the robots in (4.9) are determined through a distributed approach, based on the attraction domain of the target and local interactions among the robots. This approach is commonly used in label-free fencing studies [18–22]. Additionally, if the target position x_d is only accessible to a subset of the robots in practical scenarios, it can be communicated to all robots through a network within finite time, as discussed in [25].

Remark 4.3 Note that the internal model ϵ_i and ζ_i in Eq. (4.9) serves as a dynamic compensator to address the varying velocity of the target $v_d(t)$ which only uses position measurements, ensuring that the solution to the augmented system remains valid. Furthermore, by incorporating the inter-robot repulsion term $k_5 \sum_{k \in Ni} \alpha(|x_{i,k}|) x_{i,k}/|x_{i,k}|$ into the internal model states ϵ_i and ζ_i, such kind of design provides a critical component needed to maintain a rigid formation while fencing a target with a changing velocity, as demonstrated in Lemma 4.6. The dynamics matrices G_1 and G_2 in Eq. (4.9) can be computed from the minimal characteristic polynomial of the dynamic matrix S in (4.4), as outlined in Lemma 4.2.

Remark 4.4 In the label-free controller u_i given by (4.9), labels are merely used for identification purposes among the robots. Since every robot uses the same controller structure, swapping labels between robots does not impact the system's operation. In contrast, label-fixed designs involve controllers that set specific, predetermined relative positions between robots, resulting in unique controllers for each robot.

Next, we will demonstrate that the closed-loop MRS governed by (4.1), (4.3), (4.4), and (4.9) satisfies the following property.

P1: The collision-free rigid-formation fencing in terms of Claims (1)–(3) can be achieved by the MRS $\mathcal{V}$ (see Definition 4.1).

Moreover, we still require the following three conditions:

C1: The initial positions of the robots in $\mathcal{V}$ fulfill $\|x_{i,k}(0)\| > r,\ \forall i \neq k \in \mathcal{V}$.
C2: The parameters $k_1, k_2, k_3, k_4 \in \mathbb{R}^+$ in (4.9) satisfy

$$k_2 = k_4 \frac{k_1 + s_1 - 1}{k_3},\ \ k_1 - k_3 + s_1 - 1 > 0. \tag{4.13}$$

C3: The dynamic matrix S from (4.4) for the target is accessible to all the robots.

Remark 4.5 Here, we illustrate the meaning of the aforementioned three conditions C1–C3. Precisely, Condition C1 is essential for ensuring collision avoidance. Condition C2 simultaneously ensures the fencing property, collision avoidance, and the maintenance of a rigid formation. Condition C3 is necessary for designing the internal model in (4.9), as explained below. Each robot must know the matrix S to compute the dynamic matrices $\mathcal{G}_1$ and $\mathcal{G}_2$ for the internal model, as outlined in Lemma 4.2. Additionally, since the robots have access to the matrix S of the target, they can also determine the modal characteristics of the target's velocity v_d, such as its structure and frequency of variation (e.g., $f = 1/\sqrt{-s_1}$ Hz when $s_1 < 0$, and $f = 0$ Hz when $s_1 = 0$). This allows the robots to calculate the real-time varying velocity $v_d(t)$ based on the matrix S and the target's position $x_d(t)$. The dynamics of the target, as given in Eq. (4.3), show how $x_d(t)$ and $v_d(t)$ evolve:

$$\begin{bmatrix} x_d(t) \\ v_d(t) \end{bmatrix} = \Phi(t) \begin{bmatrix} x_d(0) \\ v_d(0) \end{bmatrix} \tag{4.14}$$

with the initial position and velocity $x_d(0)$, $v_d(0)$, and

$$\begin{aligned} \Phi(t) &= \begin{bmatrix} \cos(\sqrt{-s_1}t) & \frac{\sin(\sqrt{-s_1}t)}{\sqrt{-s_1}} \\ -\sqrt{-s_1}\sin(\sqrt{-s_1}t) & \cos(\sqrt{-s_1}t) \end{bmatrix} \otimes I_2, \text{ if } s_1 < 0, \\ \Phi(t) &= \begin{bmatrix} 1 & t \\ 0 & 1 \end{bmatrix} \otimes I_2, \text{ if } s_1 = 0. \end{aligned} \tag{4.15}$$

Here, the modal characteristics of the target velocity $v_d(t)$ are described by the structure of $\Phi(t)$ in Eq. (4.15) and its frequency, which is $f = 1/\sqrt{-s_1}$ Hz if $s_1 < 0$, and $f = 0$ Hz if $s_1 = 0$. Equations (4.14) and (4.15) show that the initial velocity $v_d(0)$ can be calculated using $\Phi(t)$, $x_d(0)$, and $x_d(t)$. This calculation allows for determining the real-time velocity $v_d(t)$ from $\Phi(t)$, $x_d(0)$, and $v_d(0)$. By considering the moving target as an exosystem, the cooperative target fencing problem in Problem 1 is transformed into a cooperative regulation problem. Condition C3 is a widely accepted assumption in the design of cooperative regulation strategies [26, 27].

To set the stage for the main technical results, we first need to cover some preliminary concepts.

Lemma 4.1 ([28]) *To lay the groundwork for the main technical results, we first consider a linear time-invariant system governed by*

$$\begin{aligned}\dot{x} &= \mathcal{A}x + \mathcal{B}u + \mathcal{E}v,\\ e &= Cx + \mathcal{D}u + \mathcal{F}v,\\ \dot{v} &= \mathcal{S}v,\end{aligned}$$

and a dynamic state feedback controller

$$u = \mathcal{K}_1 x + \mathcal{K}_2 z,\ \dot{z} = \mathcal{G}_1 z + \mathcal{G}_2 e,$$

where $x \in \mathbb{R}^n$, $u \in \mathbb{R}^m$, $e \in \mathbb{R}^p$, $z \in \mathbb{R}^{n_z}$, and $v \in \mathbb{R}^q$ represent the system state, input, regulated output, internal model state, and exosystem signal, respectively. The matrices $\mathcal{A} \in \mathbb{R}^{n\times n}$, $\mathcal{B} \in \mathbb{R}^{n\times m}$, $C \in \mathbb{R}^{p\times n}$, $\mathcal{D} \in \mathbb{R}^{p\times m}$, $\mathcal{E} \in \mathbb{R}^{n\times q}$, $\mathcal{F} \in \mathbb{R}^{p\times q}$, $\mathcal{S} \in \mathbb{R}^{q\times q}$, $\mathcal{K}_1 \in \mathbb{R}^{m\times n}$, $\mathcal{K}_2 \in \mathbb{R}^{m\times n_z}$, $\mathcal{G}_1 \in \mathbb{R}^{n_z\times n_z}$, and $\mathcal{G}_2 \in \mathbb{R}^{n_z\times p}$ are system matrices. Assume that the matrix $\mathcal{S}$ has no eigenvalues with negative real parts. If $(\mathcal{G}_1, \mathcal{G}_2)$ represents a minimal p-copy internal model of the matrix $\mathcal{S}$ and the matrix

$$\mathcal{A}_c = \begin{bmatrix} \mathcal{A} + \mathcal{B}\mathcal{K}_1 & \mathcal{B}\mathcal{K}_2 \\ \mathcal{G}_2(C + \mathcal{D}\mathcal{K}_1) & \mathcal{G}_1 + \mathcal{G}_2\mathcal{D}\mathcal{K}_2 \end{bmatrix}$$

is Hurwitz, then the Sylvester matrix equation has a unique solution $\mathcal{X} \in \mathbb{R}^{n\times q}$, $\mathcal{Z} \in \mathbb{R}^{n_z\times q}$ as

$$\begin{aligned}\mathcal{X}_c\mathcal{S} &= \mathcal{A}_c\mathcal{X}_c + \mathcal{B}_c,\\ \mathbf{0} &= C_c\mathcal{X}_c + \mathcal{D}_c\end{aligned} \tag{4.16}$$

with

$$\mathcal{X}_c = \begin{bmatrix} \mathcal{X} \\ \mathcal{Z} \end{bmatrix}, \mathcal{B}_c = \begin{bmatrix} \mathcal{E} \\ \mathcal{G}_2\mathcal{F} \end{bmatrix}, C_c = \begin{bmatrix} C\ \mathcal{D} \end{bmatrix}, \mathcal{D}_c = \begin{bmatrix} \mathcal{F} \end{bmatrix}.$$

Lemma 4.2 ([28]) *For a regulated output $e \in \mathbb{R}^p$ and an arbitrary square matrix $\mathcal{S} \in \mathbb{R}^{q\times q}$, one has that a pair of matrices $(\mathcal{G}_1, \mathcal{G}_2)$ is said to represent a minimal p-copy internal model of the matrix $\mathcal{S}$ if they are structured as follows:*

$$\mathcal{G}1 = block\ diag[\beta 1, \beta_2, \ldots, \beta_p] \in \mathbb{R}^{pq \times pq},$$
$$\mathcal{G}_2 = block\ diag[\sigma_1, \sigma_2, \ldots, \sigma_p] \in \mathbb{R}^{pq},$$

where p is the dimension of the regulated output e, block diag denotes a block diagonal matrix, $\beta_i \in \mathbb{R}^{q \times q}$ are constant square matrices, and $\sigma_i \in \mathbb{R}^q$ are constant column vectors, satisfying the following two conditions: (i) (β_i, σ_i) are controllable. (ii) The minimal characteristic polynomial of $\mathcal{S}$ divides the characteristic polynomial of β_i. Moreover, we define the minimal characteristic polynomial of $\mathcal{S}$ to be

$$S(\lambda) := \lambda^{n_m} + \alpha_1 \lambda^{n_m - 1} + \cdots + \alpha_{n_m - 1}\lambda + \alpha_{n_m},$$

which implies that $\beta_i, \sigma_i, i = 1, 2, \ldots, p$ can be designed as follows:

$$\beta_i = \begin{bmatrix} 0 & 1 & \cdots & 0 \\ 0 & 0 & \cdots & 1 \\ \vdots & \vdots & \vdots & \vdots \\ 0 & 0 & \cdots & 1 \\ -\alpha_{n_m} & -\alpha_{n_m - 1} & \cdots & -\alpha_1 \end{bmatrix}, \sigma_i = \begin{bmatrix} 0 \\ 0 \\ \cdots \\ 0 \\ 1 \end{bmatrix},$$

where the aforementioned conditions (i), (ii) of the p-copy internal model are naturally satisfied.

Remark 4.6 According to Lemma 4.2, the matrix $\mathcal{G}_1$ incorporates all the eigenvalues of the matrix $\mathcal{S}$, while the matrix $\mathcal{G}_2$ ensures that the pair $(\mathcal{G}_1, \mathcal{G}_2)$ is controllable. This configuration allows for a guaranteed solution to the Sylvester matrix equation given in Eq. (4.16) of Lemma 4.1.

Lemma 4.3 (Schur Complement [29]) *For the linear matrix inequality*

$$\begin{bmatrix} T(p) & W(p) \\ W^{\top}(p) & R(p) \end{bmatrix} > 0$$

with $T(p) = T^{\top}(p)$ and $R(p) = R^{\top}(p)$, one has that it is equivalent to the following equation:

$$1.\ T(p) > 0,\ \ R(p) - W^{\top}(p)T^{-1}(p)W(p) > 0,$$
$$2.\ R(p) > 0,\ \ T(p) - W(p)R^{-1}(p)W^{\top}(p) > 0.$$

To facilitate understanding, the main technical results are broken down into three steps: fencing, inter-robot collision avoidance, and rigid formation. We begin with a lemma addressing the fencing property in Step 1.

Lemma 4.4 *For an MRS $\mathcal{V}$ governed by* (4.1) *and* (4.9) *to collectively fence a moving target governed by* (4.3) *(i.e., satisfying Claim (1) in Eq.* (4.6)*) under condition*

C3, it is both necessary and sufficient that the control gains $k_1, k_2, k_3, k_4 \in \mathbb{R}^+$ *in* (4.9) *meet specific criteria:*

$$k_4 - s_1 k_2 - \frac{k_2^2(k_3 - s_1 k_1)}{k_1 k_2 - k_4} > 0, \quad \frac{k_1 k_2 - k_4}{k_2} > 0. \tag{4.17}$$

Proof We define $\bar{x}$ and $\bar{v}$ to be the position and velocity centers of the robot set $\mathcal{V}$, respectively,

$$\bar{x} := \frac{1}{n}\sum_{i=1}^{n} x_i, \bar{v} := \frac{1}{n}\sum_{i=1}^{n} v_i. \tag{4.18}$$

Based on the definitions of x_i, v_i in Eq. (4.1), we formulate the dynamics of $\bar{x}, \bar{v}$ in (4.18) in the Cartesian coordinates,

$$\dot{\bar{x}}(t) = \bar{v}(t), \; \dot{\bar{v}}(t) = \bar{u}(t), \tag{4.19}$$

where $\bar{u} = \frac{1}{n}\sum_{i=1}^{n} u_i$ represents the acceleration of the collective center of the robots. Given the definition of η_i in Eq. (4.11), it follows that $\sum_{i=1}^{n} \eta_i = 0$, leading to the result that

$$\bar{u} = -\left(\begin{bmatrix} k_1 & k_2 \end{bmatrix} \otimes I_2\right)\begin{bmatrix} \bar{x} \\ \bar{v} \end{bmatrix} - \left(\begin{bmatrix} k_3 & k_4 \end{bmatrix} \otimes I_2\right)\begin{bmatrix} \bar{\epsilon} \\ \bar{\zeta} \end{bmatrix} \tag{4.20}$$

where $\bar{\epsilon} := 1/n\sum_{i=1}^{n} \epsilon_i, \bar{\zeta} := 1/n\sum_{i=1}^{n} \zeta_i$ are the internal model states of the robots' centers. By defining $\varsigma := [\bar{x}^\top, \bar{v}^\top]^\top$ and $\chi := [\bar{\epsilon}^\top, \bar{\zeta}^\top]^\top$ and substituting Eq. (4.20) into Eq. (4.19), one has that the dynamic of the center of robots has the following structure:

$$\dot{\varsigma} = A\varsigma + BK_2\chi, \tag{4.21}$$

where $K_2 = [-k_3 \quad -k_4] \otimes I_2$, $B = [0 \;\; 1]^\top \otimes I_2$

$$A = \begin{bmatrix} 0 & 1 \\ -k_1 & -k_2 \end{bmatrix} \otimes I_2. \tag{4.22}$$

According to the condition $k_5\sum_{i=1}^{n} \eta_i = 0$ and $\chi = [\bar{\epsilon}^\top, \bar{\zeta}^\top]^\top$ in Eq. (4.21), it, together with (4.20) and internal model in (4.9), gives that the dynamics of χ become

$$\dot{\chi} = G_1\chi + G_2\bar{e} \tag{4.23}$$

where $\bar{e} := 1/n\sum_{i=1}^{n} e_i$ is the center of the relative position error and

$$G_1 = \begin{bmatrix} 0 & 1 \\ s_1 & 0 \end{bmatrix} \otimes I_2, G_2 = \begin{bmatrix} 0 \\ 1 \end{bmatrix} \otimes I_2. \tag{4.24}$$

Let $\sigma := [x_d^{\mathrm{T}}, v_d^{\mathrm{T}}]^{\mathrm{T}}$, the dynamics of the target, as given in Eq. (4.3) be expressed compactly as

$$\dot{\sigma} = S\sigma. \tag{4.25}$$

Following from the definitions of $e_i := x_i - x_d$ and $\varsigma := [\bar{x}^{\mathrm{T}}, \bar{v}^{\mathrm{T}}]^{\mathrm{T}}$, one has that

$$\bar{e} = \bar{x} - x_d = C\varsigma + D\sigma \tag{4.26}$$

with

$$C = \begin{bmatrix} 1 & 0 \end{bmatrix} \otimes I_2, D = \begin{bmatrix} -1 & 0 \end{bmatrix} \otimes I_2. \tag{4.27}$$

Let $\Phi := [\varsigma^{\mathrm{T}}, \chi^{\mathrm{T}}]^{\mathrm{T}}$; it, together with Eqs. (4.21), (4.23), (4.25), (4.26), gives that the augmented system of the center states is

$$\dot{\Phi} = A_c\Phi + B_c\sigma, \ \dot{\sigma} = S\sigma, \ \bar{e} = C_c\Phi + D\sigma \tag{4.28}$$

with

$$A_c = \begin{bmatrix} A & BK_2 \\ G_2C & G_1 \end{bmatrix}, B_c = \begin{bmatrix} \mathbf{0} \\ G_2D \end{bmatrix}, C_c = \begin{bmatrix} C & \mathbf{0} \end{bmatrix}.$$

According to the definitions of matrices $A, B, K_2, C, D, G_1, G_2$ in (4.21), (4.24), (4.27), one has

$$A_c = \begin{bmatrix} 0 & 1 & 0 & 0 \\ -k_1 & -k_2 & -k_3 & -k_4 \\ 0 & 0 & 0 & 1 \\ 1 & 0 & s_1 & 0 \end{bmatrix} \otimes I_2, B_c = \begin{bmatrix} 0 & 0 \\ 0 & 0 \\ 0 & 0 \\ -1 & 0 \end{bmatrix} \otimes I_2, \tag{4.29}$$

which implies that the characteristic polynomial of matrix A_c becomes $\lambda^4 + k_2\lambda_2^3 + (k_1 - s_1)\lambda^2 + (k_4 - s_1k_2)\lambda + k_3 - s_1k_1 = 0$. Given the condition of $s_1 \leq 0$, it indicates that all coefficients of the polynomial meeting $k_2 > 0, k_1 - s_1 > 0, k_4 - s_1k_2 > 0, k_3 - s_1k_1 > 0$, which further implies that A_c is Hurwitz iff

$$k_4 - s_1k_2 - \frac{k_2^2(k_3 - s_1k_1)}{k_1k_2 - k_4} > 0, \quad \frac{k_1k_2 - k_4}{k_2} > 0$$

according to Routh stability criterion [30]. Meanwhile, with the dynamic matrix S given in Eq. (4.4), the minimal characteristic polynomial of $S(\lambda)$ is calculated as $\lambda^2 - s_1 = 0$. Since the regulated output $\bar{e} := \bar{x} - x_d$ in (4.23), one has that $p = 1$. Moreover, by Lemma 4.2, one has that

$$\beta_1 = \begin{bmatrix} 0 & 1 \\ s_1 & 0 \end{bmatrix}, \sigma_1 = \begin{bmatrix} 0 \\ 1 \end{bmatrix},$$

which satisfies the Conditions (1) and (2) of the minimal p-copy internal model in Lemma 4.2. Since $\bar{x} \in \mathbb{R}^2$ and $x_d \in \mathbb{R}^2$, the matrices G_1 and G_2 of the internal model are thus expanded by the Kronecker product as follows:

$$G_1 = \begin{bmatrix} 0 & 1 \\ s_1 & 0 \end{bmatrix} \otimes I_2, G_2 = \begin{bmatrix} 0 \\ 1 \end{bmatrix} \otimes I_2,$$

which aligns with Eq. (4.24). This indicates that the matrices G_1 and G_2 in (4.24) represent a minimal p-copy internal model of the matrix S [28]. Consequently, according to Lemma 4.1, the closed-loop system described in (4.28) satisfies a Sylvester equation with a unique matrix $\mathcal{X}_c \in \mathbb{R}^{4\times 4}$ as follows:

$$\mathcal{X}_c S = A_c \mathcal{X}_c + B_c, \ \mathbf{0} = C_c \mathcal{X}_c + D \tag{4.30}$$

where A_c, B_c, C_c, D are given in (4.28). Let $\widetilde{\Phi} := \Phi - \mathcal{X}_c \sigma$ be the errors between the center states Φ and the solution states $\mathcal{X}_c \sigma$; it follows from Eqs. (4.28) and (4.30) that $\dot{\widetilde{\Phi}} = A_c \widetilde{\Phi}$, $\bar{e} = C_c \widetilde{\Phi}$, which implies that $\lim_{t\to\infty} \widetilde{\Phi}(t) = 0$, $\lim_{t\to\infty} \bar{e}(t) = 0$ because A_c is Hurwitz.

Bearing in mind (4.18) and (4.26), one has $\lim_{t\to\infty} \bar{x}(t) - x_d(t) = 1/n \sum_{i=1}^{n} x_i(t) - x_d(t) = \bar{e}(t) = 0$, which thus proves that $\lim_{t\to\infty} P_{x_d(t)}(x(t)) = 0$, i.e., the fencing property in P1. □

Remark 4.7 The condition in (4.17) solely guarantees the fencing property of an MRS $\mathcal{V}$. Moreover, this condition (4.17) can be satisfied when condition C2 is met, as will be shown in Theorem 4.1.

Lemma 4.5 *Given conditions C1 and C2, an MRS controlled by* (4.1) *and* (4.9) *ensures inter-robot collision avoidance, i.e.,* $\|x_{i,k}(t)\| > r, \forall t \geq 0, \ \forall i \neq k \in \mathcal{V}$.

Proof Let $\Phi_i := [x_i^\top, v_i^\top, \epsilon_i^\top, \zeta_i^\top]^\top$; it, together with Eqs. (4.1), (4.3), (4.9), (4.11), (4.28) gives that the derivative of Φ_i is

$$\dot{\Phi}_i = A_c \Phi_i + B_c \sigma + k_5 E \eta_i \tag{4.31}$$

where A_c, B_c in (4.29) and η_i are defined in (4.11), and $E = [0, 1, 0, -1]^\top \otimes I_2 \in \mathbb{R}^{8\times 2}$. Define $\widetilde{\Phi}_i := [\widetilde{x}_i^\top, \widetilde{v}_i^\top, \widetilde{\epsilon}_i^\top, \widetilde{\zeta}_i^\top]^\top$ to be the fencing error $\widetilde{\Phi}_i := \Phi_i - \mathcal{X}_c \sigma$; one has that the time derivative of $\widetilde{\Phi}_i$ along (4.28) and (4.31) becomes

$$\dot{\widetilde{\Phi}}_i = A_c \widetilde{\Phi}_i + (A_c \mathcal{X}_c + B_c - \mathcal{X}_c S)\sigma + k_5 E \eta_i.$$

Combining with Eq. (4.30), one has that

$$\dot{\widetilde{\Phi}}_i = A_c \widetilde{\Phi}_i + k_5 E \eta_i. \tag{4.32}$$

Equation (4.32) shows that the inter-robot repulsion η_i of robot i is present in the dynamics of $\widetilde{\Phi}_i$, affecting the convergence of the closed-loop error system (4.32).

Furthermore, the integration of the inter-robot repulsion $\sum_{i\in\mathcal{V}}\sum_{k\in Ni}\int|x_{i,k}|^R$ $\alpha(\tau)d\tau$ is introduced as a potential function within a Lyapunov candidate to ensure collision avoidance among robots. Let

$$V_p := \sum_{i\in\mathcal{V}}\sum_{k\in N_i}\int_{\|x_{i,k}\|}^{R}\alpha(\tau)d\tau \tag{4.33}$$

the integration of inter-robot repulsion be denoted for simplicity. Then, the derivative of V_p is given by

$$\begin{aligned}\frac{dV_p}{dt} &= -\sum_{i\in\mathcal{V}}\sum_{k\in N_i}\alpha(\|x_{i,k}\|)\frac{d\|x_{i,k}\|}{dt}\\ &= -\sum_{i\in\mathcal{V}}\sum_{k\in N_i}\alpha(\|x_{i,k}\|)\frac{x_{i,k}^{\mathrm{T}}}{\|x_{i,k}\|}\left(\frac{\partial x_{i,k}}{\partial x_i}\dot{x}_i+\frac{\partial x_{i,k}}{\partial x_i}\dot{x}_k\right)\\ &= -2\sum_{i\in\mathcal{V}}\sum_{k\in N_i}\alpha(\|x_{i,k}\|)\frac{x_{i,k}^{\mathrm{T}}}{\|x_{i,k}\|}\dot{x}_i.\end{aligned} \tag{4.34}$$

According to η_i given in Eq. (4.11), one has that $\sum_{i\in\mathcal{V}}\eta_i^{\mathrm{T}}v_d = v_d^{\mathrm{T}}\sum_{i\in\mathcal{V}}\eta_i = 0$, which follows from Eq. (4.34) that

$$\frac{dV_p}{dt} = -2\sum_{i\in\mathcal{V}}\eta_i^{\mathrm{T}}\dot{x}_i = -2\sum_{i\in\mathcal{V}}\eta_i^{\mathrm{T}}(\dot{x}_i - v_d) = -2\sum_{i\in\mathcal{V}}\eta_i^{\mathrm{T}}\widetilde{v}_i \tag{4.35}$$

with $\widetilde{v}_i := v_i - v_d$ and $\dot{x}_i = v_i$. Consequently, η_i is intertwined with the error state $\widetilde{v}_i$ in Eq. (4.35).

We consider a Lyapunov candidate composed of the error states $\widetilde{\Phi}_i$ and the potential function V_p, which is defined below

$$V_1(\widetilde{\Phi}_i, x_{i,k}) = \sum_{i\in\mathcal{V}}\left\{\widetilde{\Phi}_i^{\mathrm{T}}P\widetilde{\Phi}_i\right\} + \gamma k_5 V_p, \tag{4.36}$$

where $P \in \mathbb{R}^{8\times 8}$ is a positive-definite symmetric matrix, and γ is the parameter associated with P. Then, using the derivative of V_p in (4.35), the time derivative of $V_1(\widetilde{\Phi}_i, xi, k)$ along the trajectories of (4.32) becomes

$$\begin{aligned}\frac{dV_1(\widetilde{\Phi}_i, x_{i,k})}{dt} &= \sum_{i\in\mathcal{V}}\left\{\widetilde{\Phi}_i^{\mathrm{T}}(PA_c + A_c^{\mathrm{T}}P)\widetilde{\Phi}_i + k_5\left(\widetilde{\Phi}_i^{\mathrm{T}}PE\eta_i + \eta_i^{\mathrm{T}}E^{\mathrm{T}}P\widetilde{\Phi}_i\right)\right\}\\ &\quad - 2\gamma k_5\sum_{i\in\mathcal{V}}\eta_i^{\mathrm{T}}\widetilde{v}_i.\end{aligned} \tag{4.37}$$

To express the term $\widetilde{\Phi}_i^\top PE\eta_i + \eta_i^\top E^\top P\widetilde{\Phi}_i$ on the right-hand side of Eq. (4.37) in the form of $\eta_i^\top \widetilde{v}_i$, which can be canceled out by the derivative of V_p in Eq. (4.35), the matrix P is designed below

$$P = \begin{bmatrix} p_1 & p_6 & p_7 & p_6 \\ p_6 & p_2 & p_5 & p_4 \\ p_7 & p_5 & p_3 & p_5 \\ p_6 & p_4 & p_5 & p_4 \end{bmatrix} \otimes I_2 \tag{4.38}$$

where the parameters $p_i, i = 1, \ldots, 7$, are designed later. By comparing the matrices E and P in Eqs. (4.31) and (4.38), we obtain

$$k_5\left(\widetilde{\Phi}_i^\top PE\eta_i + \eta_i^\top E^\top P\widetilde{\Phi}_i\right) = 2k_5(p_2 - p_4)\widetilde{v}_i\eta_i, \tag{4.39}$$

which implies that the parameter γ in Eq. (4.36) can be set to $\gamma = p_2 - p_4$. Consequently, from Eqs. (4.37), (4.35), and (4.39), it follows that

$$\frac{dV_1(\widetilde{\Phi}_i, x_{i,k})}{dt} = \sum_{i\in\mathcal{V}}\left\{\widetilde{\Phi}_i^\top(PA_c + A_c^\top P)\widetilde{\Phi}_i\right\} = \sum_{i\in\mathcal{V}}\left\{\widetilde{\Phi}_i^\top Q\widetilde{\Phi}_i\right\} \tag{4.40}$$

with

$$Q = \begin{bmatrix} q_1 & q_5 & q_6 & q_7 \\ q_5 & q_2 & q_8 & q_9 \\ q_6 & q_8 & q_3 & q_{10} \\ q_7 & q_9 & q_{10} & q_4 \end{bmatrix} \otimes I_2, \tag{4.41}$$

$q_1 = -2k_1p_6 + 2p_6$, $q_2 = 2p_6 - 2k_2p_2, q_3 = -2k_3p_5 + 2s_1p_5$, $q_4 = -2k_4p_4 + 2p_5, q_5 = p_1 - k_2p_6 - k_1p_2 + p_4, q_6 = -k_3p_6 + s_1p_6 - k_1p_5 + p_5, q_7 = p_4 + p_7 - k_4p_6 - k_1p_4, q_8 = -k_3p_2 + s_1p_4 + p_7 - k_2p_5, q_9 = -k_4p_2 + p_5 + p_6 - k_2p_4, and q_{10} = -k_4p_5 + p_3 - k_3p_4 + s_1p_4$.

To establish the convergence of $\widetilde{v}_i$ and $\widetilde{\zeta}_i$ in Eq. (4.40), the parameters q_i for $i = 1, 3, 5, 6, 7, 8, 10$ in (4.41) can be set to 0. This leads to the following conditions: $p_5 = 0$, $p_6 = 0$, $p_1 - k_1p_2 + p_4 = 0$, $p_4 + p_7 - k_1p_4 = 0$, $-k_3p_2 + s_1p_4 + p_7 = 0$, and $p_3 - k_3p_4 + s_1p_4 = 0$. Then it derives that

$$\begin{aligned} p_1 &= \frac{k_1^2 + (s_1 - 1)k_1 - k_3}{k_3}p_4, \;\; p_2 = \frac{k_1 + s_1 - 1}{k_3}p_4, \\ p_3 &= (k_3 - s_1)p_4, \;\; p_7 = (k_1 - 1)p_4 = (k_3 - s_1)p_2, \end{aligned} \tag{4.42}$$

which implies that

$$P = \begin{bmatrix} \frac{k_1^2+(s_1-1)k_1-k_3}{k_3} & 0 & k_1-1 & 0 \\ 0 & \frac{k_1+s_1-1}{k_3} & 0 & 1 \\ k_1-1 & 0 & k_3-s_1 & 0 \\ 0 & 1 & 0 & 1 \end{bmatrix} p_4 \otimes I_2. \tag{4.43}$$

Given that $k_3 > 0$, $s_1 \leq 0$, and $p_4 > 0$, it follows that the last two diagonal terms of P satisfy $k_3 - s_1 > 0$ and $1 > 0$. Then, in accordance to Lemma 4.3, one has

$$\begin{bmatrix} \frac{k_1^2+(s_1-1)k_1-k_3}{k_3} & 0 \\ 0 & \frac{k_1+s_1-1}{k_3} \end{bmatrix} - \begin{bmatrix} \frac{(k_1-1)^2}{k_3-s_1} & 0 \\ 0 & 1 \end{bmatrix} > 0 \tag{4.44}$$

if $P > 0$. By directly calculating (4.44), we obtain

$$\begin{bmatrix} \xi_1 & 0 \\ 0 & \xi_2 \end{bmatrix} > 0 \tag{4.45}$$

with

$$\xi_1 = \frac{(s_1+1)k_1k_3 - k_3^2 - s_1k_1^2 - (s_1-1)s_1k_1 + (s_1-1)k_3}{k_3(k_3-s_1)},$$

$$\xi_2 = \frac{k_1 - k_3 + s_1 - 1}{k_3}.$$

As $k_3 > 0$, $k_3 - s_1 > 0$, we only need to consider the numerators of ξ_1 and ξ_2 to ensure that $P > 0$.

Case 1: If $s_1 = 0$, it is derived that $(k_1 - k_3 + s_1 - 1)/k_3 > 0$ for $P > 0$.

Case 2: If $s_1 < 0$, rewrite the numerator of ξ_1 in (4.45) in descending order of k_1 as

$$f(k_1) = -s_1k_1^2 + \big((s_1+1)k_3 - (s_1-1)s_1\big)k_1 - k_3^2 + (s_1-1)k_3,$$

which implies that $f(k_1)$ is a quadratic function of k_1. From the fact that

$$\begin{aligned} k_3 + 1 - s_1 - \frac{(s_1+1)k_3 - (s_1-1)s_1}{2s_1} &= \frac{2s_1k_3 + 2s_1 - 2s_1^2 - (s_1+1)k_3 + (s_1-1)s_1}{2s_1} \\ &= \frac{s_1k_3 + s_1 - s_1^2 - k_3}{2s_1} \\ &= \frac{(s_1-1)(k_3-s_1)}{2s_1} > 0, \end{aligned}$$

it can be observed that $f(k_1)$ increases monotonically when $k_1 > k_3 + 1 - s_1$. Substituting $k_1 = k_3 + 1 - s_1$ into $f(k_1)$ yields

$$f(k_1) = -s_1(k_3+1-s_1)^2 + \big((s_1+1)k_3 - (s_1-1)s_1\big) \\ \times (k_3+1-s_1) - k_3^2 + (s_1-1)k_3 = 0,$$

which implies that $f(k_1) > 0$ (i.e., $P > 0$) if $k_1 > k_3 + 1 - s_1$.

Accordingly, it concludes that $\xi_1 > 0, \xi_2 > 0$ (i.e., $P > 0$) if $(k_1 - k_3 + s_1 - 1)/k_3 > 0$. Note that $k_1^2 + (s_1-1)k_1 - k_3 = k_1(k_1+s_1-1) - k_3 > 0$ with $k_1 > 1$ and $k_1 - 1 - k_3 + s_1 > 0$ in (4.43). Moreover, the condition $k_1 + s_1 - 1 > k_3$ implies $p_2 - p_4 > 0$ with (4.42), which guarantees V_1 in Eq. (4.36) is positive definite.

Next, it follows from the fact $p_5 = 0$, $p_6 = 0$, and Eq. (4.42) that

$$\frac{dV_1(\widetilde{\Phi}_i, x_{i,k})}{dt} = \sum_{i\in\mathcal{V}} \left\{ \begin{bmatrix} \widetilde{v}_i \\ \widetilde{\zeta}_i \end{bmatrix}^{\mathrm{T}} \bar{Q} \begin{bmatrix} \widetilde{v}_i \\ \widetilde{\zeta}_i \end{bmatrix} \right\} \tag{4.46}$$

with

$$\bar{Q} = \begin{bmatrix} -2k_2p_2 & -k_4p_2 - k_2p_4 \\ -k_4p_2 - k_2p_4 & -2k_4p_4 \end{bmatrix} \otimes I_2.$$

Since the leading principal minors of $\bar{Q}$ satisfy $-2k_2p_2 < 0$ and

$$-2k_2p_2 \times -2k_4p_4 - (-k_4p_2 - k_2p_4)^2 = -(k_4p_2 - k_2p_4)^2,$$

it follows from Eq. (4.42) that $k_4p_2 - k_2p_4 = 0$ if

$$k_4\frac{k_1+s_1-1}{k_3} = k_2, \tag{4.47}$$

which then implies that the matrix $\bar{Q}$ is negative semidefinite. It follows from Eq. (4.42) that

$$k_4\frac{k_1+s_1-1}{k_3} = k_2. \tag{4.48}$$

Combining the condition $k_1 > k_3 + 1 - s_1$ and Eq. (4.47) together gives C2. Moreover, one has that

$$\begin{aligned} \frac{dV_1(\widetilde{\Phi}_i, x_{i,k})}{dt} &= \sum_{i\in\mathcal{V}} \{-2(\frac{k_1+s_1-1}{k_3})^2 k_4p_4\widetilde{v}_i^2 - 2k_4p_4(\widetilde{\varepsilon}_i)^2 - 4\frac{k_1+s_1-1}{k_3}k_4p_4\widetilde{v}_i\widetilde{\varepsilon}_i\} \\ &= -\sum_{i\in\mathcal{V}} \frac{2p_4}{k_4}(\frac{(k_1+s_1-1)k_4}{k_3}\widetilde{v}_i^2 + k_4\widetilde{\varepsilon}_i)^2 \\ &= -\sum_{i\in\mathcal{V}} \frac{2p_4}{k_4}\|(k_2\widetilde{v}_i + k_4\widetilde{\zeta}_i)\|^2 \le 0. \end{aligned} \tag{4.49}$$

To simplify, let $V_1(t) := V_1(\widetilde{\Phi}i(t), xi, k(t))$ denote the function V_1 at time t. It then follows from Eq. (4.49) that

$$V_1(\overline{T}) = V_1(0) + \int_0^{\overline{T}} \frac{dV_1(\widetilde{\Phi}_i, x_{i,k})}{dt} dt \leq V_1(0) \tag{4.50}$$

for an arbitrary constant time $\overline{T} > 0$. Combining this with the definition of V_1 in Eq. (4.36), we obtain that

$$k_5(p_2 - p_4) \sum_{i \in \mathcal{V}} \sum_{k \in \mathcal{N}_i} \int_{\|x_{i,k}(\overline{T})\|}^{R} \alpha(s) ds \leq V_1(\overline{T}) \leq V_1(0). \tag{4.51}$$

Under condition C1, one has that $V_1(0)$ is bounded, so is $\sum_{i \in \mathcal{V}} \sum_{k \in \mathcal{N}_i} \int_{\|x_{i,k}(\overline{T})\|}^{R} \alpha(s) ds$, $\forall \overline{T} \geq 0$ in Eq. (4.51). However, using the fact

$$\lim_{\|x_{i,k}(\overline{T})\| \to r^+} \sum_{i \in \mathcal{V}} \sum_{k \in \mathcal{N}_i} \int_{\|x_{i,k}(\overline{T})\|}^{R} \alpha(s) ds = \infty, \tag{4.52}$$

where α is defined in (4.12) and r^+ represents the right-hand limit of r, we conclude that $|x_{i,k}(\overline{T})| > r$ for all $i \in \mathcal{V}, i, k \in \mathcal{V}$, and $\overline{T} > 0$. This proves collision avoidance. □

Remark 4.8 Lemma 4.5 aims to demonstrate inter-robot collision avoidance through the boundedness of the Lyapunov function V_1 in (4.36), which is composed of two terms. The first term relates to the error states $\widetilde{\Phi}_i$ between robot i and the target, while the second term represents the integration of inter-robot repulsion, V_p. These two terms in V_1 cannot converge to zero but only to an invariant set of partial states, $k_2\widetilde{v}_i + k_4\widetilde{\zeta}_i = 0$, due to (4.49). Moreover, if both terms were to converge to zero, it would imply that all robot positions coincide, resulting in $V_1 = \infty$, which contradicts the condition that $dV_1(t)/dt \leq 0$ in (4.49). It is also noteworthy that the chosen function V_1 and the invariant set $k_2\widetilde{v}_i + k_4\widetilde{\zeta}_i = 0$ contribute to the rigid formation described in Lemma 4.6.

4.4 Rigid Fencing Pattern and Analysis

Lemma 4.6 *Under conditions C1–C3, an MRS governed by* (4.1) *and* (4.9) *achieves a rigid formation while moving with the target described by* (4.3), *i.e.,*

$$\lim_{t \to \infty} (v_i(t) - v_d(t)) = 0.$$

Proof From Eq. (4.49) in the proof of Lemma 4.5, we see that $dV_1/dt = 0$ occurs if and only if $k_2\widetilde{v}_i + k_4\widetilde{\zeta}_i = 0$. This implies that the largest invariant set $\{\widetilde{x}_i, \widetilde{v}_i, \widetilde{\epsilon}_i, \widetilde{\zeta}_i \mid dV_1/dt = 0\}$ consists of the line defined by $k_2\widetilde{v}_i + k_4\widetilde{\zeta}_i = 0$. Since any line segment between two points on this line remains on the line, the invariant set is compact. By the LaSalle invariant set theorem [31], it follows that the trajectories of $\widetilde{x}_i, \widetilde{v}_i, \widetilde{\epsilon}_i, \widetilde{\zeta}_i$ converge to

$$\lim_{t\to\infty} k_2\widetilde{v}_i(t) + k_4\widetilde{\zeta}_i(t) = 0. \tag{4.53}$$

Then, we will demonstrate that

$$\text{If } k_2\widetilde{v}_i + k_4\widetilde{\zeta}_i = 0, \text{ then } v_i(t) - v_d(t) = 0, \forall i \in \mathcal{V}. \tag{4.54}$$

According to the dynamics of $\dot{\widetilde{v}}_i, \dot{\widetilde{\zeta}}_i$ in (4.32), one has $\dot{\widetilde{\zeta}}_i = \widetilde{x}_i + s_1\widetilde{\epsilon}_i - k_5\eta_i$. Taking the derivative of $\dot{\widetilde{v}}_i$ yields that

$$\begin{aligned}
\ddot{\widetilde{v}}_i =& -k_2\dot{\widetilde{v}}_i - k_4\dot{\widetilde{\varepsilon}}_i - (k_1 - 1)\widetilde{v}_i - (k_3 - s_1)\widetilde{\varepsilon}_i - \ddot{\widetilde{\varepsilon}}_i \\
=& -(k_2\dot{\widetilde{v}}_i + k_4\dot{\widetilde{\zeta}}_i) - \left(k_1 - 1 - \frac{(k_3 - s_1)k_2}{k_4}\right)\widetilde{v}_i - (k_3 - s_1)\left(\frac{k_2}{k_4}\widetilde{v}_i + \widetilde{\zeta}_i\right) - \ddot{\widetilde{\zeta}}_i.
\end{aligned} \tag{4.55}$$

Subtracting $k_2/k_4\ddot{\widetilde{v}}_i$ at both sides of (4.55), it then follows from Eq. (4.48) that

$$\ddot{\widetilde{v}}_i = s_1\widetilde{v}_i + \frac{k_4}{k_2 - k_4}(k_2\dot{\widetilde{v}}_i + k_4\dot{\widetilde{\zeta}}_i) + \frac{(k_3 - s_1)}{k_2 - k_4}(k_2\widetilde{v}_i + k_4\widetilde{\zeta}_i) + \frac{1}{k_2 - k_4}(k_2\ddot{\widetilde{v}}_i + k_4\ddot{\widetilde{\zeta}}_i). \tag{4.56}$$

Utilizing the fact that $\dot{\widetilde{v}}_i$, $\dot{\widetilde{\zeta}}_i$, $\ddot{\widetilde{v}}_i$, and $\ddot{\widetilde{\zeta}}_i$ are uniformly continuous, it follows from (4.54) that

$$k_2\dot{\widetilde{v}}_i + k_4\dot{\widetilde{\zeta}}_i = 0, \; k_2\ddot{\widetilde{v}}_i + k_4\ddot{\widetilde{\zeta}}_i = 0. \tag{4.57}$$

Comparing (4.54), (4.56), (4.57) gives

$$\ddot{\widetilde{v}}_i = s_1\widetilde{v}_i, \tag{4.58}$$

and one has that $\ddot{\widetilde{\zeta}}_i = s_1\widetilde{\zeta}_i$. From the condition of $\dot{\widetilde{\epsilon}}_i = \widetilde{\zeta}_i$, it gives that

$$\dot{\widetilde{\zeta}}_i(t) = s_1\widetilde{\epsilon}_i(t) - m_i \tag{4.59}$$

with a suitable constant vector $m_i \in \mathbb{R}^2$, and considering the fact that $\dot{\widetilde{\zeta}}_i = s_1\widetilde{\epsilon}_i(t) + \widetilde{x}_i - k_5\eta_i$ from (4.32), it follows that

$$\widetilde{x}_i(t) - k_5 \sum_{k \in \mathcal{N}_i} \left\{ \alpha(\|x_{i,k}\|) \frac{x_{i,k}}{\|x_{i,k}\|} \right\} = m_i, \forall i \in \mathcal{V}. \tag{4.60}$$

Case 1: When $s_1 = 0$, the problem simplifies to a fencing problem with a constant-velocity target (see, e.g., [22]). According to Eq. (4.58), we have $\lim_{t\to\infty} \dddot{\widetilde{v}}_i(t) = 0$. Therefore, it is sufficient to show by contradiction that $\lim_{t\to\infty} \dot{\widetilde{v}}_i(t) = 0$ and $\lim_{t\to\infty} \widetilde{v}_i(t) = 0$, which follows a similar approach to [22]. Thus, we conclude that $\lim_{t\to\infty} \widetilde{v}_i(t) = v_i(t) - v_d(t) = 0$ for all $i \in \mathcal{V}$.

Case 2: If $s_1 < 0$, it follows from Eq. (4.58) and the differential equation from [32] that the solution for $\widetilde{v}_i$ is given by

$$\widetilde{v}_i(t) = c_{i,1} \cos(\sqrt{-s_1}t) + c_{i,2} \sin(\sqrt{-s_1}t) \tag{4.61}$$

where $c_{i,1}$ and $c_{i,2}$ are parameters determined by the initial states. Next, we will prove by contradiction that $c_{i,1} = 0$ and $c_{i,2} = 0$. It follows from Eqs. (4.32), (4.61) that $\widetilde{x}_i$

$$\widetilde{x}_i(t) = \frac{c_{i,1}}{\sqrt{-s_1}} \sin(\sqrt{-s_1}t) - \frac{c_{i,2}}{\sqrt{-s_1}} \cos(\sqrt{-s_1}t) + d_i \tag{4.62}$$

with a constant vector $d_i \in \mathbb{R}^2$. It can be deduced that $\widetilde{x}_i$ in (4.62) is a periodic function if $c_{i,1} \neq 0$ or $c_{i,2} \neq 0$, which contradicts with the condition of (4.60). Then, it derives that $c_{i,1} = 0$, $c_{i,2} = 0$, $\forall i \in \mathcal{V}$, which implies that $\widetilde{v}_i(t) = v_i(t) - v_d(t) = 0$ with (4.61) (i.e., the proof of (4.54) is thus completed).

Since the term $k_2\widetilde{v}_i + k_4\widetilde{\zeta}_i$ is uniformly continuous, according to (4.54), for any $\delta_1 > 0$, there exists $\delta_2 > 0$ such that

$$\left\| k_2\widetilde{v}_i + k_4\widetilde{\zeta}_i \right\| < \delta_2, \ \forall i \in \mathcal{V},$$

which leads to the conclusion that

$$\left\| v_i - v_d \right\| < \delta_1, \forall i \in \mathcal{V}.$$

Since $\lim_{t\to\infty}(k_2\widetilde{v}_i(t) + k_4\widetilde{\zeta}_i(t)) = 0$ as given in (4.53), there exists a constant $T > 0$ such that for all $t \geq T$, $|k_2\widetilde{v}_i + k_4\widetilde{\zeta}_i| < \delta_2$ for all $i \in \mathcal{V}$. This implies that $|v_i - v_d| < \delta_1$ for all $i \in \mathcal{V}$. Consequently, we conclude that $\lim t \to \infty(v_i(t) - v_d(t)) = 0$ for all $i \in \mathcal{V}$, indicating a rigid formation. This completes the proof. □

With Lemmas 4.4–4.6, it is ready to present the main technical results.

Theorem 4.1 *Under Conditions C1, C2, and C3, an MRS $\mathcal{V}$, governed by* (4.1) *and* (4.9), *and interacting with a moving target described by* (4.3), *satisfies property P1. Therefore, Problem 1 is successfully addressed.*

Proof Firstly, Lemma 4.4 demonstrates that the fencing property is achieved if the control gains k_1, k_2, k_3, and k_4 satisfy the condition given in (4.17).

Secondly, according to the collision avoidance and rigid-formation properties established in Lemmas 4.5 and 4.6, the control gains must also satisfy Condition C2 from (4.63), specifically

$$k_2 = k_4 \frac{k_1 + s_1 - 1}{k_3} > 0, \quad k_1 - k_3 - 1 + s_1 > 0. \tag{4.63}$$

Substituting (4.63) into (4.17) from Lemma 4.4, one has that

$$k_4 - s_1 k_2 - \frac{k_2^2(k_3 - s_1 k_1)}{k_1 k_2 - k_4} = \frac{(k_1 + s_1 - 1 - k_3)k_4^2}{k_3(k_1 k_2 - k_4)} > 0,$$
$$\frac{k_1 k_2 - k_4}{k_2} = \frac{k_1 + s_1 - 1 - k_3}{k_1 + s_1 - 1} > 0,$$

which implies that condition C2 is sufficient to satisfy (4.17) (i.e., Remark 4.7). Therefore, Problem 1 is solved under conditions C1–C3, completing the proof. □

Remark 4.9 In contrast to label-fixed strategies, the label-free approach outlined in (4.9) can be applied directly without additional calculations or design adjustments when increasing the number of robots. Furthermore, this label-free design offers greater flexibility in varying environments and specific scenarios, facilitating a more straightforward achievement of fencing formation. Additionally, it is more robust in the event of robot failures. An illustrative example will be presented in the following section.

4.5 Simulation Results

In this section, we consider an MRS governed by (4.1), (4.9) with $n = 4$ robots and a moving target described by (4.3) with a variable velocity. We set the sensing range and safe distance to $R = 10$ and $r = 2$, respectively. Accordingly, the potential function $\alpha(\cdot)$ in (4.12) is defined with these values: $r = 2$ and $R = 10$.

Additionally, we set $s_1 = -0.1$ to ensure the target moves periodically, starting from the initial position $x_d(0) = [2, 8]^\top$ and with the initial velocity $v_d(0) = [0.5, 0.5]^\top$. The target's trajectories are shown by the red dashed curves in Fig. 4.3. The parameters in (4.9) are chosen as $k_1 = 2.2$, $k_2 = 6$, $k_3 = 0.1$, $k_4 = 3$, and $k_5 = 20$, ensuring that condition C2 is met. Figure 4.3a–d depicts the evolution of robots from various initial states (blue circles), meeting condition C1, toward their final rigid-formation states (yellow circles) while tracking the moving target (green square) with $P_{x_d}(x) = 0$. These figures illustrate that the robots successfully achieve fencing formation from any starting point, independent of their labels. To demonstrate the advantages of the label-free strategy, Fig. 4.4 compares this with simulations using a label-fixed approach, starting from the same initial positions as Fig. 4.3. In this approach, specific desired relative positions for each robot with respect to the

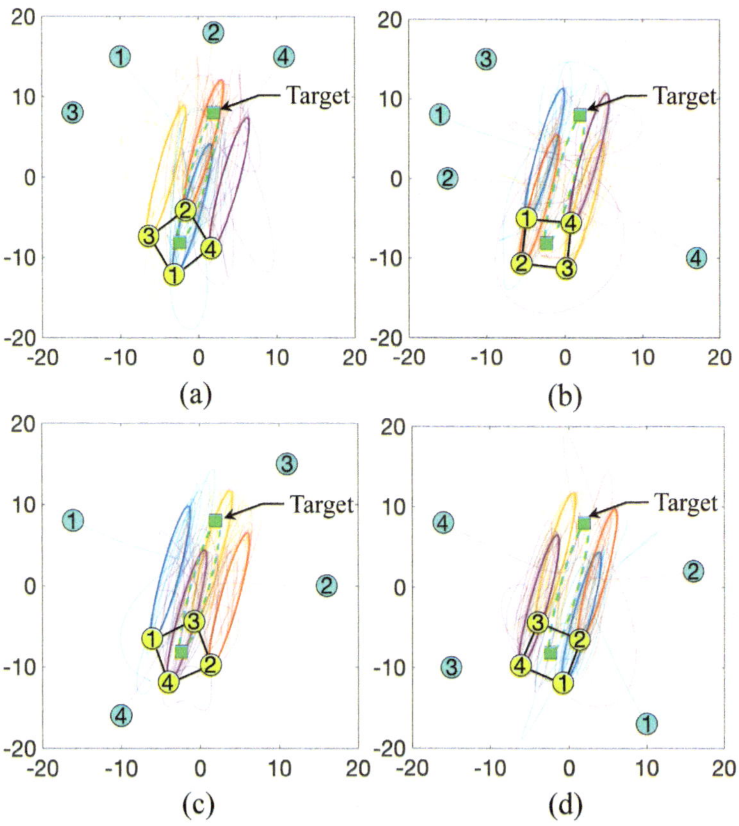

Fig. 4.3 a–d Trajectories of the robots from random initial positions and velocities to collision-free rigid-formation target fencing with the proposed label-free controller (4.9) (here, the blue and yellow circles denote the initial and final states of the robots, respectively, whereas the green square the target)

target are pre-set as $[-7, -7]^\top$, $[7, -7]^\top$, $[7, 7]^\top$, and $[-7, 7]^\top$. The results shown in Fig. 4.4a–d reveal that, with fixed labels and relative positions, the robots achieve rigid-formation fencing from their initial states (blue circles) to their final positions (yellow circles).

Comparing Fig. 4.3a–d with Fig. 4.4a–d, it is evident that the label-fixed approach results in more oscillations during the fencing process, as shown in Fig. 4.4. This highlights the label-free method's greater flexibility and efficiency in minimizing the robots' movement. Figure 4.5a demonstrates the robustness of the label-free fencing controller (4.9). When robot 4 fails at $t = 20$ s, the remaining robots 1, 2, and 3 successfully maintain a triangular formation to continue fencing the moving target. In contrast, the label-fixed strategy fails to maintain the formation under the same conditions, as shown in Fig. 4.5b. This confirms the label-free design's robustness in handling robot breakdowns.

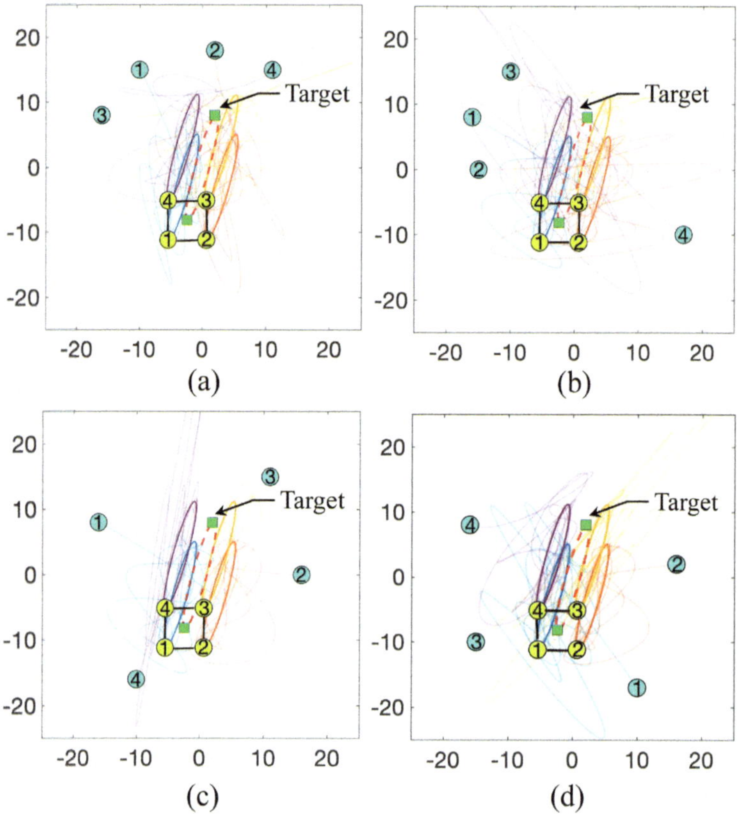

Fig. 4.4 **a–d** Trajectories of the robots from the same initial positions and velocities in Fig. 4.3 to collision-free rigid-formation target fencing with a label-fixed strategy (here, the blue and yellow circles denote the initial and final states of the robots, respectively, whereas the green square the target)

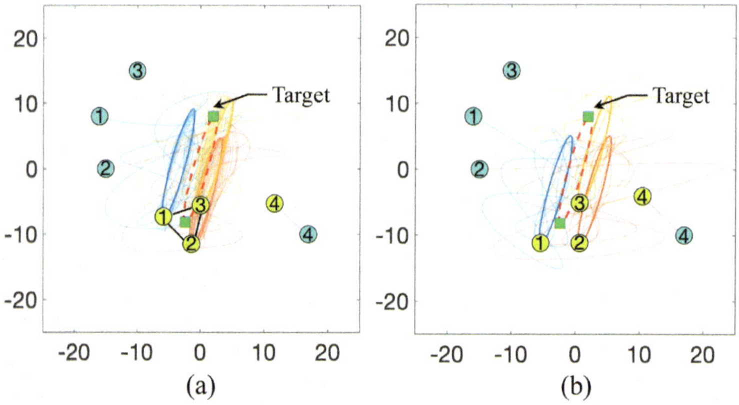

Fig. 4.5 A special case when robot 4 suddenly breaks down at $t = 20$ s. Trajectory comparison of the robots and the target between the proposed label-free fencing controller (4.9) (see subfigure (**a**)) and a label-fixed strategy (see subfigure (**b**))

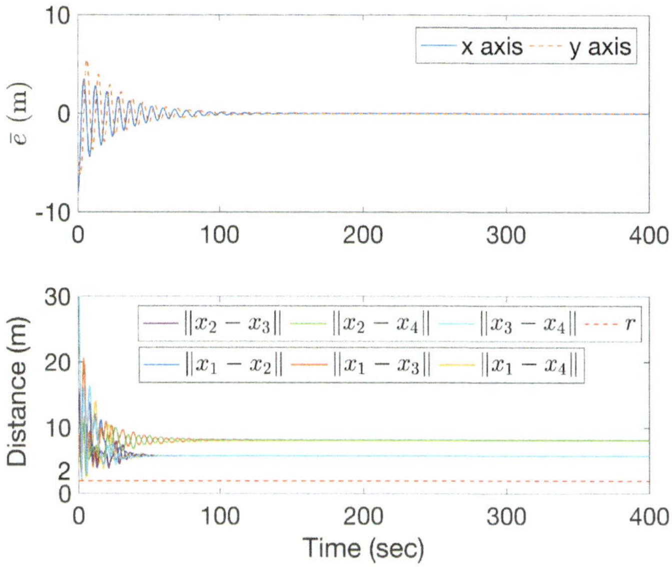

Fig. 4.6 Temporal evolution of the fencing error $\bar{e}$ (see Eq. (4.26)) and the relative distances $\|x_i - x_j\|, i \neq k, i, k \in \mathcal{V}$ among robots in Fig. 4.3b for example

In the label-free fencing simulation depicted in Fig. 4.3b, we examine the state evolution. Figure 4.6 shows the state evolution where $\lim_{t\to\infty} \bar{e}(t) = 0$ achieves $P_{x_d}(x) = 0$ implicitly in P1. The pairwise distances between robots remain greater than 2, ensuring inter-robot collision avoidance with $r = 2$. To compare the label-free approach quantitatively, Fig. 4.7 presents the state evolution for the label-fixed strategy, corresponding to Fig. 4.4b. Both approaches converge to zero fencing errors within 110 s, indicating similar efficiency in achieving the fencing goal. However, the pairwise distance oscillations in Fig. 4.7b are 60% larger than those in Fig. 4.6b for the label-free strategy. Additionally, the convergence time for the distances in the label-fixed strategy is 50 s longer compared to the label-free approach. Figure 4.8 illustrates that $\lim_{t\to\infty} x_i(t) - x_d(t) = d_i \neq 0$ and $\lim_{t\to\infty} v_i(t) - v_d(t) = 0$ for the simulation in Fig. 4.3b, confirming that the rigid formation described in Definition 4.1 is achieved. This demonstrates the feasibility of Theorem 4.1 and verifies the advantages of the label-free fencing controller.

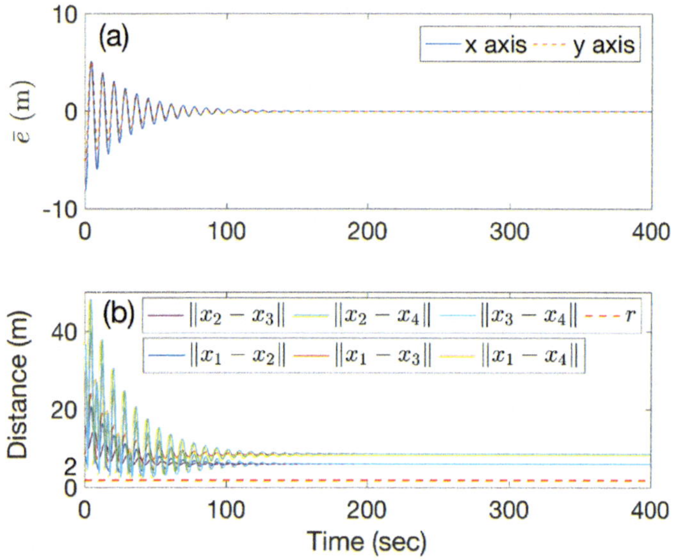

Fig. 4.7 Temporal evolution of the fencing error $\bar{e}$ (see Eq. (4.26)) and the relative distances $\|x_i - x_j\|, i \neq k, i, k \in \mathcal{V}$ among robots in Fig. 4.4b for example

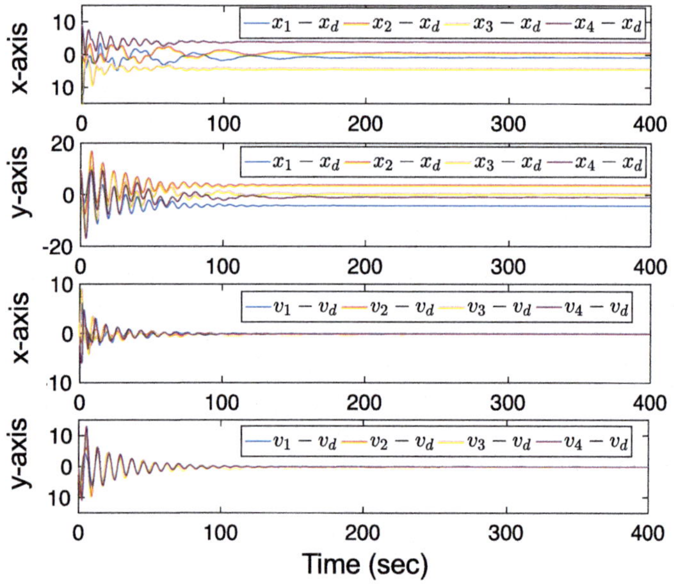

Fig. 4.8 Temporal evolution of the position error $x_i - x_d$ and velocity error $v_i - v_d$, $i = 1, 2, 3, 4$, in Fig. 4.3b for example

4.6 Conclusion

In this chapter, we have introduced a label-free control strategy that enables second-order MRSs to cooperatively fence a moving target with variable velocity within a convex hull. This approach ensures both inter-robot collision avoidance and rigid formation without the need for pre-assigned labels for specific robots. The effectiveness of the proposed cooperative control protocol has been demonstrated through numerical simulations.

References

1. R. Olfati-Saber, Flocking for multi-agent dynamic systems: algorithms and theory. IEEE Trans. Autom. Control **51**(3), 401–420 (2006)
2. H.-T. Zhang, C. Zhai, Z. Chen, A general alignment repulsion algorithm for flocking of multi-agent systems. IEEE Trans. Autom. Control **56**(2), 430–435 (2010)
3. Z. Chen, H.-T. Zhang, A remark on collective circular motion of heterogeneous multi-agents. Automatica **49**(5), 1236–1241 (2013)
4. X. Jin, Y. Shi, Y. Tang, X. Wu, Event-triggered attitude consensus with absolute and relative attitude measurements. Automatica **122**, 109245 (2020)
5. Z. Meng, W. Ren, Z. You, Distributed finite-time attitude containment control for multiple rigid bodies. Automatica **46**(12), 2092–2099 (2010)
6. Y. Yang, D. Constantinescu, Y. Shi, Passive multiuser teleoperation of a multirobot system with connectivity-preserving containment. IEEE Trans. Robot. **38**(1), 209–228 (2021)
7. Z. Chen, H.-T. Zhang, No-beacon collective circular motion of jointly connected multi-agents. Automatica **47**(9), 1929–1937 (2011)
8. Z. Sun, H.G. de Marina, G.S. Seyboth, B.D.O. Anderson, C. Yu, Circular formation control of multiple unicycle-type agents with nonidentical constant speeds. IEEE Trans. Control Syst. Technol. **27**(1), 192–205 (2018)
9. Y. Lan, G. Yan, Z. Lin, Distributed control of cooperative target enclosing based on reachability and invariance analysis. Syst. Control Lett. **59**(7), 381–389 (2010)
10. T.-H. Kim, T. Sugie, Cooperative control for target-capturing task based on a cyclic pursuit strategy. Automatica **43**(8), 1426–1431 (2007)
11. R. Zheng, Y. Liu, D. Sun, Enclosing a target by nonholonomic mobile robots with bearing-only measurements. Automatica **53**, 400–407 (2015)
12. X. Yu, L. Liu, Distributed circular formation control of ring-networked nonholonomic vehicles. Automatica **68**, 92–99 (2016)
13. F. Chen, W. Ren, Y. Cao, Surrounding control in cooperative agent networks. Syst. Control Lett. **59**(11), 704–712 (2010)
14. Y.J. Shi, R. Li, K.L. Teo, Cooperative enclosing control for multiple moving targets by a group of agents. Int. J. Control **88**(1), 80–89 (2015)
15. B.-B. Hu, H.-T. Zhang, J. Wang, Multiple-target surrounding and collision avoidance with second-order nonlinear multi-agent systems. IEEE Trans. Ind. Electron. **68**(8), 7454–7463 (2021)
16. B.-B. Hu, H.-T. Zhang, B. Liu, H. Meng, G. Chen, Distributed surrounding control of multiple unmanned surface vessels with varying interconnection topologies. IEEE Trans. Control Syst. Technol. **30**(1), 400–407 (2021)
17. K. Sakurama, H.-S. Ahn, Multi-agent coordination over local indexes via clique-based distributed assignment. Automatica **112**, 108670 (2020)

18. C. Wang, G. Xie, Limit-cycle-based decoupled design of circle formation control with collision avoidance for anonymous agents in a plane. IEEE Trans. Autom. Control **62**(12), 6560–6567 (2017)
19. B. Liu, Z. Chen, H. Zhang, X. Wang, T. Geng, H. Su, J. Zhao, Collective Dynamics and Control for Multiple Unmanned Surface Vessels. IEEE Trans. Control Syst. Technol. **28**(6), 2540–2547 (2020)
20. Z. Chen, A cooperative target-fencing protocol of multiple vehicles. Automatica **107**, 591–594 (2019)
21. L. Kou, Z. Chen, J. Xiang, Cooperative fencing control of multiple vehicles for a moving target with an unknown velocity. IEEE Trans. Autom. Control **67**(2), 1008–1015 (2021)
22. L. Kou, Y. Huang, Z. Chen, S. He, J. Xiang, Cooperative fencing control of multiple second-order vehicles for a moving target with and without velocity measurements. Int. J. Robust Nonlinear Control **31**(10), 4602–4615 (2021)
23. B. Hu, Z. Chen, H. Zhang, Distributed moving target fencing in a regular polygon formation. IEEE Trans. Control Netw. Syst. **9**(1), 210–218 (2021)
24. B. Hu, H. Zhang, Bearing-only motional target-surrounding control for multiple unmanned surface vessels. IEEE Trans. Ind. Electron. **69**(4), 3988–3997 (2021)
25. Z. Sun, S. Mou, M. Deghat, B.D. Anderson, A.S. Morse, Finite time distance-based rigid formation stabilization and flocking. IFAC Proc. Vol. **47**(3), 9183–9189 (2014)
26. Y. Su, J. Huang, Cooperative output regulation of linear multi-agent systems. IEEE Trans. Autom. Control **57**(4), 1062–1066 (2011)
27. Su, Y., Huang, J.: Cooperative output regulation with application to multi-agent consensus under switching network. IEEE Trans. Syst. Man Cybern. Part B Cybern. **42**(3), 864–875 (2012)
28. J. Huang, *Nonlinear Output Regulation: Theory and Applications* (SIAM, Philadelphia, USA, 2004)
29. S. Boyd, L. El Ghaoui, E. Feron, V. Balakrishnan, *Linear Matrix Inequalities in System and Control Theory* (SIAM, Philadelphia, USA, 1994)
30. E.J. Routh, *A Treatise on the Stability of a Given State of Motion: Particularly Steady Motion* (Macmillan and Company, London, England, 1877)
31. K.H. Khalil, *Nonlinear Systems* (Prentice Hall, Upper Saddle River, 2002)
32. L. Hörmander, Hypoelliptic second order differential equations. Acta Math. **119**(1), 147–171 (1967)

Chapter 5
Distributed Fencing Control via Long-Term Task Execution

5.1 Introduction

The previous Chaps. 3 and 4 introduced complex fencing with an evenly rotating formation and a varying-velocity target in 2D space, however, they still lack a formal method to develop the fencing control with general dynamics and even higher-dimensional Euclidean space. To this end, in this chapter, we propose a long-term task term algorithm (LTTE) algorithm for a multi-robot system to fence a moving target into an *ordering-flexible* convex hull resiliently in the n-dimensional Euclidean space. In what follows, we will introduce the related background.

Over the years, the concept of *multi-robot target fencing*, where a group of robots collectively provides protection for a target, has gained increasing importance in complex urban scenarios, such as highway vehicle conveyance [1], marine vessel escort [2], and multi-UAV protection [3]. In these applications, the primary challenge lies in designing a coordinated fencing strategy that allows the robots to collectively form a convex hull, effectively ensuring that the target remains secured within the interior of the formation.

A straightforward and convenient approach to forming such a convex hull involves predefining the desired robot-target relative positions (or displacements) with a fixed spatial ordering, known as ordering-fixed target fencing. Early studies focused on achieving fixed-order fencing formations for static targets [4], which was later extended to moving targets [5, 6]. For more complex scenarios involving multiple moving targets, a distributed fencing controller was introduced in [7], relying on the estimation of the targets' center. However, these ordering-fixed approaches [4–7] inherently require predesignated positions and a fixed ordering of robots within the controller, limiting their flexibility in dynamic environments. This lack of adaptability can lead to inefficiencies in energy usage and even deadlock situations.

Alternatively, a more adaptable and efficient method known as the *ordering-flexible* strategy has gained attention in recent studies. Unlike the ordering-fixed

B.-B. Hu and H.-T. Zhang, *Multi-Robot Systems*, Cognitive Intelligence and Robotics,
https://doi.org/10.1007/978-981-97-9214-6_5

methods [4–7], this approach enables target fencing within a convex hull without predefining the spatial order of robots [8]. The robot positions naturally emerge through interactions during fencing rather than being rigidly assigned beforehand, thereby addressing issues like deadlock and inefficiency. Initial efforts focused on this concept include an output-regulation algorithm for static target fencing [9] and a limit-cycle-based structure that adds collision avoidance to static target scenarios [10]. These methods were later expanded to constant-velocity targets in an ordering-flexible manner [11–13]. A dynamic regulator incorporating an internal model was developed for targets with variable velocities to maintain a rigid fencing formation [14]. However, while these studies [8–14] have made significant strides in realizing ordering-flexible fencing, they still fall short of providing a comprehensive framework that guarantees the feasibility of this approach in dynamically changing conditions. Moreover, these methods have primarily been applied in 2D environments, leaving the challenge of higher-dimensional ordering-flexible target fencing in more complex scenarios largely unexplored.

Drawing inspiration from the long-term task execution frameworks in [15, 16], this chapter addresses the challenge of ensuring the feasibility of *ordering-flexible* fencing by integrating it into an online constraint-based optimization strategy. The proposed LTTE algorithm enables a multi-robot system to effectively fence a moving target within an *ordering-flexible* convex hull in n-dimensional Euclidean space, even in dynamic environments. The approach involves defining the target-approaching and sensing-neighbor collision-avoidance subtasks as constraints within the optimization process, ensuring continuous adaptability during long-term operations. Slack variables are incorporated to manage potential conflicts between these constraints, balancing the attraction forces driving the robots toward the target and the repulsion forces from collision-avoidance requirements. This mechanism allows the robots to naturally form a fencing configuration with flexible spatial orderings. The efficacy, adaptability across dimensions, robustness, and obstacle avoidance of the proposed LTTE algorithm are demonstrated through 2D experiments and 3D simulations. The main contributions of this chapter are three-fold.

1. Propose an LTTE algorithm to build an online constraint-based optimization framework that enables multi-robot *long-term* target fencing with arbitrary spatial orderings in evolving environments.
2. Introduce slack variables to accommodate the relaxation of various subtask constraints and ensure rigorous asymptotic convergence to *ordering-flexible* target fencing despite the strong nonlinear interactions imposed by time-varying collision-avoidance constraints.
3. Showcase the effectiveness of the proposed method with 2D *ordering-flexible* target-fencing experiments involving three AMRs. Additionally, validate its multi-dimensional adaptability, robustness, and obstacle avoidance capabilities through 3D simulations addressing various initial positions, sudden robot breakdowns, and static obstacles.

The structure of this chapter is as follows: Sect. 5.2 introduces the definitions and formulations for long-term task execution and *ordering-flexible* target fencing. Section 5.3 covers the analysis of asymptotic convergence. Section 5.4 presents the results of 2D experiments and 3D simulations. Finally, Sect. 5.5 summarizes the conclusions.

5.2 General Fencing Problem Formulation

5.2.1 Long-Term Task Execution

Assume a task set $\mathcal{T}$ is characterized by an implicit function $\phi(\sigma)$ [17]

$$\mathcal{T} := \{\sigma \in \mathbb{R}^d \mid \phi(\sigma) \leq 0\}, \tag{5.1}$$

where $\sigma = [\sigma_1, \cdots, \sigma_d]^\top \in \mathbb{R}^d$ represents the dummy coordinates with $d \in \mathbb{R}^+$ indicating the dimensionality, and $\phi(\cdot) : \mathbb{R}^d \to \mathbb{R}$ is a twice continuously differentiable function (i.e., $\phi(\cdot) \in C^2$). The condition $\phi(\sigma) \leq 0$ in Eq. (5.1) defines the region of the target set $\mathcal{T}$, encompassing all possible states within $\mathcal{T}$.

Remark 5.1 The formulation in Eq. (5.1) enables describing the target set $\mathcal{T}$ as either a point or a 1D (or 2D) manifold region, depending on the selection of $\phi(\sigma)$. As illustrated in Fig. 5.1, for example, a point-target set can be defined by $\phi(\sigma, p_0) := |\sigma - p_0|$ with $p_0 = [0.5, 0.5, 0.5]^\top$, while a target ball is defined by $\phi(\sigma) := \sigma_1^2 +$

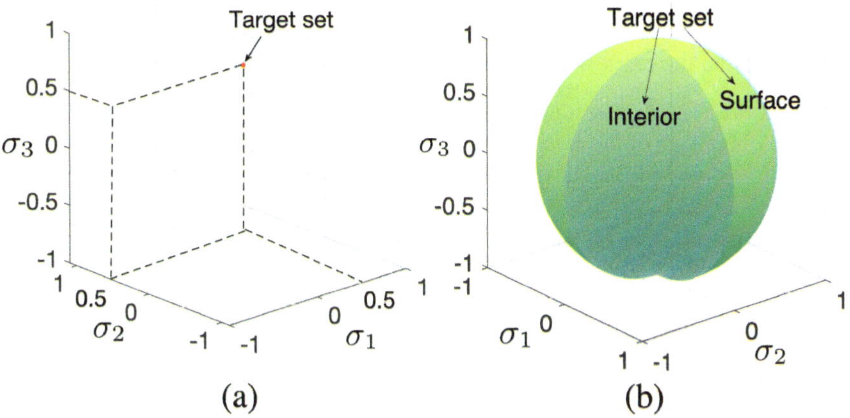

Fig. 5.1 **a** A point-target set $\mathcal{T} : \phi(\sigma) := (\sigma_1 - 0.5)^2 + (\sigma_2 - 0.5)^2 + (\sigma_3 - 0.5)^2 \leq 0$ only contains a red point $p_0 = [0.5, 0.5, 0.5]^\top$. **b** A ball-target set $\mathcal{T} : \phi(\sigma) = \sigma_1^2 + \sigma_2^2 + \sigma_3^2 - 1 \leq 0$ contains both the surface and interior of the green ball

$\sigma_2^2 + \sigma_3^2 - r^2$ with radius $r = 1$. Instead of using the complex distance calculation $\text{dist}(p_1, \mathcal{T}) := \inf |p - p_1| \mid p \in \mathcal{T}$ to measure the distance between a point $p_1 \in \mathbb{R}^d$ and the target set $\mathcal{T}$, $\phi(p_1)$ provides a simpler implicit method for measuring this distance. For instance, with the target ball $\phi(\sigma) := \sigma_1^2 + \sigma_2^2 + \sigma_3^2 - r^2$, the value $\phi(p_1)$ is positive if the point p_1 is outside the ball and $\phi(p_1) \leq 0$ if p_1 is on or within the ball. However, there are some pathological cases where this implicit method might not be sufficient, such as when $\lim_{t\to\infty} \phi(p_1(t)) \leq 0$ does not necessarily imply that $\lim_{t\to\infty} \text{dist}(p_1(t), \mathcal{T}) = 0$. To address these issues, the following assumption will be introduced.

Assumption 5.1 ([17]) For any given $\kappa > 0$ and a point $p_1(t) \in \mathbb{R}^d$, it follows that $\inf \phi(p_1(t)) : \text{dist}(p_1(t), \mathcal{T}) \geq \kappa > 0$.

Assumption 5.1 guarantees that the implicit function $\phi(p_1(t))$ can be used to accurately represent the distance from the point p_1 to the target set $\mathcal{T}$. This condition can be fulfilled by certain polynomial functions [17].

Defining $x \in \mathbb{R}^d$ to represent the state vector of the robot and substituting x into the parameterization of $\mathcal{T}$ in Eq. (5.1), one has that $\phi(x)$ provides an implicit measure of the distance between the state x and the target set $\mathcal{T}$. Specifically, x is outside the set $\mathcal{T}$ if $\phi(x) > 0$, and x is within the set $\mathcal{T}$ if $\phi(x) \leq 0$. When x is not in the set $\mathcal{T}$ (i.e., $\phi(x) > 0$), we can introduce the minimization of $\phi(x)$ to ensure that x converges to $\mathcal{T}$: Then, when x is not in the set $\mathcal{T}$ (i.e., $\phi(x) > 0$), we are ready to introduce the minimization of $\phi(x)$ to make x converge to $\mathcal{T}$,

$$\min_u \phi(x) \text{ s.t. } \dot{x} = f(x) + g(x)u \tag{5.2}$$

where $u \in \mathbb{R}^q$ is the control input, and $f(\cdot) : \mathbb{R}^d \to \mathbb{R}^d$ and $g(\cdot) : \mathbb{R}^d \to \mathbb{R}^{d\times q}$ are locally Lipschitz continuous functions. The constraint $\dot{x} = f(x) + g(x)u$ in Eq. (5.2) represents a standard control affine dynamic structure that applies to a broad range of robots [18]. Essentially, $\phi(x)$ decreases along the optimal input u in Eq. (5.2), indicating that the robot will eventually approach the target set $\mathcal{T}$.

Remark 5.2 The dimension of the state vector $d \in \mathbb{R}^+$ in Eq. (5.2) can be greater than the dimension of the robot's operational space $n \in \mathbb{R}^+$ (i.e., $d \geq n$). This allows $x \in \mathbb{R}^d$ to include additional states, such as orientation, linear velocities, and angular velocities of UAVs [19]. However, for this chapter, this extended representation is less pertinent since Eq. (5.2) primarily offers a general optimization approach for ensuring that the state vector x converges to the target set $\mathcal{T}$. In the context of the *ordering-flexible* target fencing problem discussed in this chapter, our focus is specifically on the n-dimensional Euclidean space where the robot operates, rather than the extended state space. This is because the position information $x_i \in \mathbb{R}^n$, governed by the single-integrator dynamics in Eq. (5.5), suffices for formulating the position-based subtasks outlined in Eqs. (5.9) and (5.12). The issue of underactuated dynamics is addressed by external low-level tracking modules, such as output regulation control and sliding mode control [9], which are beyond the scope of this chapter. For details on orientation regulation, please see Remark 5.10 for details.

Remark 5.3 At each time step, Eq. (5.2) is solved to find the optimal input u^*. In this framework, $\phi(x)$ is treated as a cost function, and u^* can be determined using a standard gradient-descent method [20]. The optimal input is computed as $u^* = g^+(-f(x) - \partial\phi(x)/\partial x)$, where $g^+ \in \mathbb{R}^{q\times d}$ denotes the Moore–Penrose inverse of $g(x)$, satisfying $gg^+ = I_d$, with $I_d \in \mathbb{R}^{d\times d}$ representing the d-dimensional identity matrix. By differentiating $\phi(x)$ with respect to time t and substituting u^*, we get

$$\dot{\phi}(x) = \frac{\partial\phi(x)}{\partial x^\top}\dot{x} = \frac{\partial\phi(x)}{\partial x^\top}\Big(f(x) + g(x)g^+\Big(-f(x) - \frac{\partial\phi(x)}{\partial x}\Big)\Big) = -\left\|\frac{\partial\phi(x)}{\partial x}\right\|^2 \leq 0.$$

In this case, the function $\phi(x)$ will explicitly decrease along the optimal trajectories as long as $|\partial\phi(x)/\partial x| \neq 0$. However, when $\phi(x)$ is not a convex function, convergence to local minima is a typical outcome that cannot be avoided.

However, the traditional task execution in (5.2) only tries the best to make the states x approach the target set $\mathcal{T}$ because the function $\phi(x)$ is minimized as a cost function, which may fail to work any longer in dynamic environments (i.e., $x(T_1) \in \mathcal{T} \nRightarrow x(t) \in \mathcal{T}, \forall t > T_1 > 0$). To rigorously guarantee the task performance for long-term autonomy, a recent work [16] has proposed an approach of long-term task execution by incorporating $\phi(x)$ into the constraint rather than the cost function.

Definition 5.1 (*Long-term task execution* [15]) For a target set $\mathcal{T}$ in Eq. (5.1), a robot governed by Eq. (5.2) achieves the long-term task execution if the following constraint-based optimization is solved, i.e.,

$$\min_{u,\delta} \|u\|^2 + |\delta|^2 \text{ s.t. } L_f h(x) + L_g h(x)u \geq -\gamma(h(x)) - \delta, \tag{5.3}$$

where $h(x) = -\phi(x)$ satisfying the inequality in (5.3) is a control barrier function (CBF) [22], $L_f h(x)$, $L_g h(x)$ denote the Lie derivatives of $h(x)$ along the functions f and g in Eq. (5.2), respectively, and $\delta \geq 0$ represents a slack variable to measure the violation extent of execution to target set $\mathcal{T}$. Here, $\gamma(\cdot) : (-b, a) \to (-\infty, +\infty)$ for some $a, b > 0$, referred to as the extended class $\mathcal{K}$ function, is strictly increasing, and satisfies $\gamma(0) = 0$, which can determine how fast the states x approaching the target set $\mathcal{T}$ by setting $\gamma(h(x)) := h(x)^Q$ with different odd positive integer Q [21].

In Definition 5.1, the long-term task execution (5.3) ensures robust task performance in dynamic environments by explicitly constraining the state evolution of $h(x) := -\phi(x)$. Fundamentally, this long-term behavior is guaranteed because the CBF $h(x)$ provides resilient forward invariance of the states x with respect to the target set $\mathcal{T}$ in a constrained manner [22], meaning that

$$x(T_1) \in \mathcal{T} \Rightarrow x(t) \in \mathcal{T}, \forall t > T_1 > 0. \tag{5.4}$$

5.2.2 Ordering-Flexible Target Fencing

In this subsection, we consider a multi-robot system consisting of N robots, labeled as $\mathcal{V} = 1, 2, \ldots, N(N \geq 3)$, where each robot follows a single-integrator dynamic model [17]:

$$\dot{x}i = u_i, \quad i \in \mathcal{V}, \tag{5.5}$$

with $x_i \in \mathbb{R}^n$ and $u_i \in \mathbb{R}^n$ representing the position and control input of robot i in the n-dimensional Euclidean space, respectively. The control input u_i is bounded by a common value $\zeta \in \mathbb{R}^+$, satisfying $|u_i|\infty \leq \zeta$. For broader applicability, the input u_i in Eq. (5.5) can be interpreted as the high-level desired guidance velocity, making it relevant for robots with higher-order dynamics. This model can be extended to various platforms, including UAVs, AGVs, and USVs [23–28].

Assumption 5.2 We assume that the low-level velocity-tracking module can asymptotically track the high-level desired velocity u_i^* of robot $i, i \in \mathcal{V}$, in Eq. (5.5), which implies that $\lim_{t\to\infty} v_i(t) = u_i^*$, where v_i is the actual velocity of robot i.

Since the ROS controllers[1] embedded in many commercially available robotic systems are typically simple PID controllers that only achieve asymptotic convergence, Assumption 5.2 is both practical and necessary.

Then, the neighbor set of robot i is defined as follows:

$$\mathcal{N}_i := \{j \in \mathcal{V}, j \neq i, \mid \|x_{i,j}(t)\| \leq R\}, \tag{5.6}$$

where the collision radius $r \in \mathbb{R}^+$ and the sensing radius $R \in \mathbb{R}^+$ share the same center, where $R > r$. The relative position between robots i and j is denoted as $x_{i,j} := x_i - x_j$. Since $|x_{i,j}(t)|$ varies with time, the robots in the neighbor set $\mathcal{N}_i$ are also time-varying. This dynamic nature facilitates the *ordering-flexible* fencing but presents challenges in the subsequent convergence analysis.

We also consider a moving target governed by the following dynamics,

$$\dot{x}_d = v_d, \tag{5.7}$$

where $x_d \in \mathbb{R}^n$ and $v_d \in \mathbb{R}^n$ denote the position and velocity of the target. According to Eq. (5.7), the target may move with either a constant velocity (i.e., $\dot{v}_d = 0$) or a non-constant velocity (i.e., $\dot{v}_d \neq 0$). It is important to note that $|v_d|_\infty < \zeta$; otherwise, the robots will be unable to keep up with the moving target.

Assumption 5.3 We assume that the robots have access only to the target's position $x_d(t)$, not to its velocity v_d. Therefore, each robot $i \in \mathcal{V}$ has a local velocity estimator $\widehat{v}_d^i$, where $\widehat{v}_d^i, i \in \mathcal{V}, \ \|\widehat{v}_d^i\|_\infty < \zeta$, such that $\lim_{t\to\infty}\{\widehat{v}_d^i(t) - v_d(t)\} = \mathbf{0}_n$ exponentially.

---

[1] ROS controller: https://slaterobotics.medium.com/how-to-implement-ros-control-on-a-custom-robot-748b52751f2e.

Remark 5.4 Assumption 5.3 outlines the practical feasibility of estimating the moving target's velocity v_d from position-only measurements, assuming each robot is equipped with position sensors. While the design of such local estimators $\widehat{v}_d^i$ is well-documented in the literature, such as through output regulation theory [29–31], these specific methods are beyond the scope of this article. Nonetheless, it's important to note that the vanishing estimation errors $\widehat{v}_d^i - v_d$ can impact the convergence of the *ordering-flexible* target fencing task. To illustrate, we provide an example of a local estimator for the case where $\dot{v}_d = 0$ below.

$$\dot{\widehat{x}}_d^i = -\chi_1(\widehat{x}_d^i - x_d) + \widehat{v}_d^i, \ \dot{\widehat{v}}_d^i = -\chi_1\chi_2(\widehat{x}_d^i - x_d), \tag{5.8}$$

with the ith estimator $\widehat{\Omega}_i = [\widehat{x}_d^i, \widehat{v}_d^i]^\top$ and the estimation gains $\chi_1 \in \mathbb{R}^+$ and $\chi_2 \in \mathbb{R}^+$, let $\widetilde{\Omega}_i = \widehat{\Omega}_i - [x_d, v_d]^\top$. It follows that $\dot{\widetilde{\Omega}}_i = A\widetilde{\Omega}_i$, where A is a Hurwitz matrix below

$$A = \begin{bmatrix} -\chi_1 & 1 \\ -\chi_1\chi_2 & 0 \end{bmatrix},$$

which thus implies $\lim_{t\to\infty} \widetilde{\Omega}_i(t) = \mathbf{0}_{2n}$, i.e., $\lim_{t\to\infty}\{\widehat{v}_d^i(t) - v_d(t)\} = \mathbf{0}_n$.

To define *ordering-flexible* target fencing, we first introduce the spatial ordering sequence $s[i]$, for $i \in \mathbb{Z}_1^N$, of robots arranged within a convex hull. Different rules can determine this spatial ordering. For instance, in 2D, one approach is to start with the smallest relative angle between each robot and the centroid of all robots, and then establish the ordering sequence $s[1], \ldots, s[N]$ in an anticlockwise direction around the centroid. This is illustrated in Fig. 5.2: in subfigure (a), the ordering sequence is $s[1] = 2, s[2] = 6, s[3] = 1, s[4] = 3, s[5] = 4, s[6] = 5$, and in subfigure (b), it is $s[1] = 5, s[2] = 3, s[3] = 2, s[4] = 4, s[5] = 6, s[6] = 1$. For 3D coordination, analogous ordering rules can be applied.

Definition 5.2 (*Ordering-flexible target fencing*) For a multi-robot system $\mathcal{V}$ governed by Eq. (5.5) to achieve *ordering-flexible* fencing for a moving target (5.7), the following three objectives must be fulfilled,

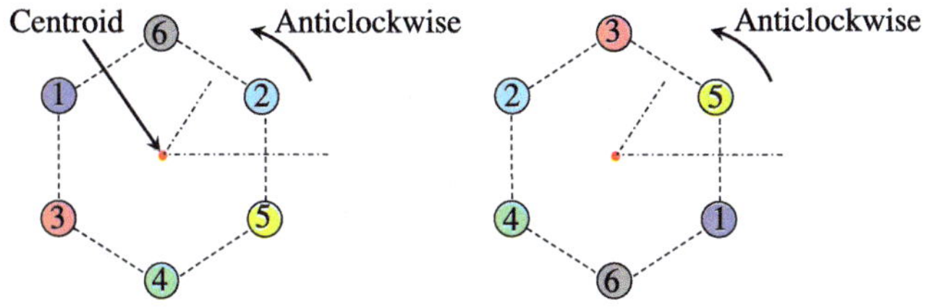

(a) Ordering: {2-6-1-3-4-5} (b) Ordering: {5-3-2-4-6-1}

Fig. 5.2 Two spatial ordering sequences of a 6-robot hexagonal formation in 2D. (The circles represent the robots, and the red node is the centroid)

1. **(Convex-hull fencing)** The moving target is fenced into the interior of the convex hull formed by all the robots, i.e.,

$$\lim_{t\to\infty}\left\{\sum_{i\in\mathcal{V}} x_i(t)/N - x_d(t)\right\} = \mathbf{0}_n$$

where x_i is the position of robot i, and N is the number of robots.

2. **(Flexible-ordering formation)** The robots' fencing formation converges to a stable region with a consistent spatial ordering, i.e.,

$$\begin{aligned}&(a)\ \lim_{t\to\infty}\left\{\dot{x}_{s[i]}(t) - \dot{x}_{s[i+1]}(t)\right\} = \mathbf{0}_n,\\ &(b)\ r \le \lim_{t\to\infty}\left\|x_{s[i]}(t) - x_{s[i+1]}(t)\right\| < R,\\ &\forall i \in \mathbb{Z}_1^N\ (\text{If } i = N, \text{then } s[i+1] = s[1]),\end{aligned}$$

where $s[1], s[2], \ldots, s[N]$ denote the spatial ordering sequence of the robots in the convex hull, calculated by a bijection $s[i] : \mathbb{Z}_1^N \to \mathbb{Z}1^N$ from the robot labels $\mathcal{V} = 1, \ldots, N$; $x_{s[i]}$ represents the position of the $s[i]$th robot; and R is the sensing radius as defined in (5.6).

3. **(Collision & Overlapping avoidance)** The avoidance of both inter-robot collisions and robot-target overlaps is ensured simultaneously, i.e., $\|x_{i,j}(t)\| \ge r$, $\|x_{i,d}(t)\| > 0, \forall i \ne j \in \mathcal{V}, t > 0$ with $x_{i,j}$ given in Eq. (5.6) and $x_{i,d} := x_i - x_d$ the relative position between robot i and the target.

In Definition 5.2, Objective (1) ensures the fundamental requirement of target fencing. Condition (a) in Objective (2) describes the implicit rigid formation, while condition (b) highlights the concept of flexible-ordering formation by maintaining a steady ordering according to a specified rule. Notably, no specific spatial ordering is predetermined for the robots. Objective (3) addresses the prevention of undesirable collisions and overlapping between robots and the target, thereby ensuring that the robots eventually form the desired convex hull. Illustrative examples of this are provided in Fig. 5.3, where circle robots with varying orderings fence a triangular target in both 2D and 3D Euclidean spaces.

5.2.3 Encoding Long-Term Target Fencing

To encode the long-term *ordering-flexible* target fencing, we design two subtasks based on Definitions 5.1 and 5.2: sensing-neighbor collision avoidance and target approaching.

From Assumption 5.1, it follows that the target approaching subtask $\mathcal{T}_{i,0}$ for robot i is described as

$$\phi_{i,0}(x_i) = \|x_i - x_d\|, i \in \mathcal{V}. \tag{5.9}$$

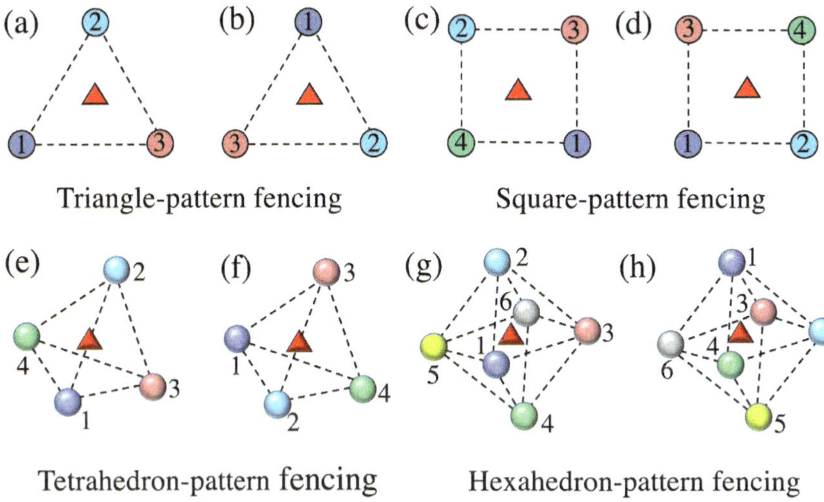

Fig. 5.3 Illustration of four kinds of *ordering-flexible* target fencing in 2D and 3D. (The circles in different colors represent the robots, and the red triangle is the target)

It can be seen that $\phi_{i,0}(x_i)$ in (5.9) involves only the point x_d from the target set $\mathcal{T}_0$ and is twice differentiable, provided that $|x_{i,d}(t)| \neq 0, \forall t > 0$, a condition which Lemma 5.2 will ensure later. Thus, given the long-term execution framework in Eq. (5.3) and the dynamics in Eqs. (5.5) and (5.7), the derivative of $\phi_{i,0}(x_i)$ in Eq. (5.9) satisfies

$$\frac{\partial \phi_{i,0}(x_i)}{\partial x_i^{\mathsf{T}}}(u_i - \widehat{v}_d^i) \leq -\gamma_1(\phi_{i,0}(x_i)) + \delta_{i,0}, \tag{5.10}$$

where $\delta_{i,0} \geq 0$ represents the slack variable, $\widehat{v}_d^i$ is the ith local estimator for target's velocity v_d in Assumption 5.3, and $\gamma_1(\psi_{i,0}(x_i))$ is designed to be

$$\gamma_1\big(\phi_{i,0}(x_i)\big) = \eta_1 \|x_{i,d}\| \tag{5.11}$$

with the parameter $\eta_1 \in \mathbb{R}^+$ and $x_{i,d} = x_i - x_d$ as given in Definition 5.2, Eq. (5.11) includes η_1, an arbitrary positive constant that dictates the approach speed of the robot toward the target. Increasing η_1 leads to a faster approach speed u_i for the robot, up to the maximum input norm ζ, i.e., $|u_i|_\infty \leq \zeta$ in (5.15d). The simple formulation of the function $\gamma_1(\cdot)$ in Eq. (5.11) will be used in Lemma 5.3 to demonstrate convex-hull fencing.

Similarly, the sensing-neighbor collision-free subtask $\mathcal{T}_{i,j}$ for robot i, where $j \in \mathcal{N}_i$, is designed to be

$$\phi_{i,j}(x_i) = r - \|x_{i,j}\|, i \in \mathcal{V}, j \in \mathcal{N}_i, \tag{5.12}$$

where $\mathcal{N}_i$, the specified collision radius $r \in (0, R)$, and $x_{i,j}$ are defined in (5.6). It is observed that the target set in Eq. (5.12) is positioned outside the circle, i.e., $|x_{i,j}| > r$, which aligns with the collision avoidance definition. Furthermore, Moreover, $\phi_{i,j}(x_i)$ is also twice differentiable for $\|x_{i,j}(t)\| \neq 0, \forall t > 0$, which will be guaranteed in Lemma 5.2 as well.

Note that the radius r is stipulated to be larger than actual distance for inter-robot collision avoidance, which will be utilized to eliminate the influence of target-attracting force in Eq. (5.9) later. Moreover, from Definition 5.1, one has that $\phi_{i,j}(x_i)$, $j \in \mathcal{N}_i$, in Eq. (5.12) satisfy

$$\frac{\partial \phi_{i,j}(x_i)}{\partial x_i^{\top}}(u_i - \widehat{v}_d^i) \leq -\gamma_2(\phi_{i,j}(x_i)), \tag{5.13}$$

where $\gamma_2(\phi_{i,j}(x_i))$ is another kind of extended class $\mathcal{K}$ function

$$\gamma_2\big(\phi_{i,j}(x_i)\big) = \frac{2\zeta\phi_{i,j}(x_i)}{\eta_2} \tag{5.14}$$

where $\eta_2 \in \mathbb{R}^+$ is a constant, and ζ is given in (5.5). The function $\gamma_2(\phi_{i,j}(x_i))$ in Eq. (5.14) is designed using ζ and will be used to demonstrate that collision-free subtasks for non-neighboring robots are inherently met, as shown in Lemma 5.1. The constant $\eta_2 \in \mathbb{R}^+$ in Eq. (5.14) determines the minimum value of the sensing radius R in Assumption 5.4. Specifically, increasing η_2 results in a larger sensing radius R. Conversely, if η_2 is too small, the collision-free constraints for non-neighboring robots in Lemma 5.1 might not be satisfied.

Inspired by Eq. (5.3) for long-term task execution, multi-robot long-term target fencing is formulated as a decentralized constraint-based optimization problem for robot i. This formulation is derived from Eqs. (5.9), (5.10), (5.12), and (5.13).

$$\min_{u_i, \delta_{i,0}} \left\{ \|u_i - \widehat{v}_d^i\|^2 + l\delta_{i,0}^2 \right\} \tag{5.15a}$$

$$\text{s.t. } \frac{\partial \phi_{i,0}}{\partial x_i^{\top}}(u_i - \widehat{v}_d^i) + \gamma_1(\phi_{i,0}) - \delta_{i,0} \leq 0, \tag{5.15b}$$

$$\frac{\partial \phi_{i,j}}{\partial x_i^{\top}}(u_i - \widehat{v}_d^i) + \gamma_2(\phi_{i,j}) \leq 0,\ j \in \mathcal{N}_i, \tag{5.15c}$$

$$\|u_i\|_\infty \leq \zeta, \forall i \in \mathcal{V}, \tag{5.15d}$$

where the position x_i, the input u_i, the ith velocity estimator $\widehat{v}_d^i$, and the slack variable $\delta i, 0 \in \mathbb{R}^+$ are specified in Eqs. (5.5) and (5.10), while $l \in \mathbb{R}^+$ denotes the corresponding weight. For clarity, the functions $\phi_{i,0} = \phi_{i,0}(x_i)$, $\phi_{i,j} = \phi_{i,j}(x_i)$, $\gamma_1(\phi_{i,0}) = \gamma_1(\phi_{i,0}(x_i))$, and $\gamma_2(\phi_{i,j}) = \gamma_2(\phi_{i,j}(x_i))$ are given in Eqs. (5.10), (5.13), (5.11), and (5.14), respectively.

The cost function in (5.15a) aims to minimize two objectives simultaneously. The minimization of $|u_i - \widehat{v}_d^i|^2$ focuses on tracking the target's velocity, while minimizing

$\delta_{i,0}^2$ aims to reduce the distance between robot i and the target x_d. The weight l is set to $l > 1$, indicating that reducing the robot-target distance is prioritized over velocity tracking. The constraint in (5.15b) represents the CBF constraint for performing the target-approaching subtask $\mathcal{T}_{i,0}$ for robot i. The inclusion of $\delta_{i,0}$ ensures the feasibility of this constraint at the initial time, given that $x_i(0)$ is not in the target set $\mathcal{T}_{i,0}$ (i.e., $x_i(0) \notin \mathcal{T}_{i,0}$), otherwise, the constraint might not be satisfied initially. The constraint in (5.15c) represents the CBF constraint for ensuring collision avoidance between robots i and j, where $j \in \mathcal{N}_i$. The absence of slack variables in this constraint guarantees that $|x_{i,j}| \geq r$ will be strictly maintained. Finally, the constraint in (5.15d) ensures that the control input remains within the bounded limit.

Remark 5.5 To ensure the feasibility of Eq. (5.15), a feasible solution always exists, which can be expressed as

$$\{u_i = \widehat{v}_d^i, \delta_{i,0} > 0 \text{ is sufficiently large}\}. \tag{5.16}$$

This solution can satisfy all the constraints (5.15b)–(5.15d). Specifically, for the solution provided in Eq. (5.16), constraint (5.15b) is satisfied when $\delta_{i,0}$ is sufficiently large. From the condition of $|x_{i,j}(0)| \geq r$, as detailed in Assumption 5.5 later, one has that $\gamma_2(\phi_{i,j}(0)) \leq 0$. This ensures that constraints (5.15c) are satisfied due to the forward-invariance property discussed in Lemma 5.2 later. Additionally, constraint (5.15d) is naturally satisfied since $|\widehat{v}d^i|_\infty < \zeta$, as shown in Assumption 5.3.

Remark 5.6 Similar to the LTTE algorithm in Definition 5.1, the proposed framework in Eq. (5.15) focuses on minimizing the energy consumption associated with tracking the target's velocity within the cost function. Furthermore, by incorporating time-varying target-fencing subtasks into the rigorous constraints (5.15b) and (5.15c), the long-term property of Eq. (5.15) accommodates ordering flexibility to address varying environmental factors. These factors include different initial positions, potential sudden breakdowns of some robots, and static obstacles, which will be illustrated through experiments and simulations in Sect. 5.4 later.

Now, we are ready to introduce the main problem addressed in this chapter.

Problem 1 (*Long-term target fencing*) We design a coordinated optimized controller

$$\begin{aligned}[u_i^\mathsf{T}, \delta_{i,0}]^\mathsf{T} := \varpi\big(&\phi_{i,0}, \gamma_1(\phi_{i,0}),\\ &\phi_{i,j}, \ldots, \gamma_2(\phi_{i,j})\big), i \in \mathcal{V}, j \in \mathcal{N}_i,\end{aligned} \tag{5.17}$$

for achieving multi-robot long-term *ordering-flexible* target fencing in changing environments, the constraint-based optimization problem (5.15) is utilized. Here, $\varpi(\cdot) : \mathbb{R} \to \mathbb{R}^{n+1}$ represents an unknown implicit function.

5.3 Long-Term Task Execution Controller Design and Convergence Analysis

In this section, we present the technical results related to Objectives 1–3 of *ordering-flexible* fencing as defined in Definition 5.2. This is necessary because these three objectives are not directly intuitive from the long-term target fencing described in (5.15).

Before proceeding with the detailed analysis, we first need to ensure that collision-free constraints between non-neighboring robots (i.e., $j \notin \mathcal{N}_i \mid |x_{i,j}| > R$) are satisfied. Although these constraints are not explicitly shown in (5.15), they are crucial due to the discontinuity of $\partial\phi_{i,j}/\partial x_i^\top$ when $|x_{i,j}| > R$.

Assumption 5.4 The sensing radius R is chosen to be at least as large as a specified value, i.e., $R \geq r + \eta_2$, where r is given in Eq. (5.12) and $\eta_2 \in \mathbb{R}^+$ is a positive parameter.

Assumption 5.4 specifies a minimal sensing radius that will be used to ensure the collision-free constraints for non-neighboring robots, as demonstrated in Lemma 5.1.

Lemma 5.1 *Under Assumption 5.4, all robots $\mathcal{V}$ governed by (5.5) naturally satisfy the collision-free constraints (5.13) for non-neighboring robots with arbitrary input, i.e.,*

$$-\frac{x_{i,j}^\top}{\|x_{i,j}\|}(u_i - \widehat{v}_d^i) \leq -\gamma_j(\phi_{i,j}), \forall \|u_i\|_\infty \leq \zeta, \|\widehat{v}_d^i\|_\infty \leq \zeta, i \in \mathcal{V}, j \notin \mathcal{N}_i,$$

with $\eta_2 \in \mathbb{R}^+$ given in Eq. (5.14).

Proof For arbitrary non-neighboring robots $i \neq j \in \mathcal{V}$ satisfying $|x_{i,j}| > R$, it follows from Assumption 5.4 and the definition of $\phi_{i,j}$ in Eq. (5.12) that

$$\phi_{i,j} = r - \|x_{i,j}\| \leq -\eta_2 < 0. \tag{5.18}$$

Using $\gamma_2(\phi_{i,j})$ in Eq. (5.14), it follows from Eq. (5.18) that

$$\gamma_2(\phi_{i,j}) \leq -2\zeta. \tag{5.19}$$

Meanwhile, the constraints in Eq. (5.13) satisfy

$$\frac{\partial\phi_{i,j}}{\partial x_i^\top}(u_i - \widehat{v}_d^i) = -\frac{x_{i,j}^\top}{\|x_{i,j}\|}(u_i - \widehat{v}_d^i) \leq 2\zeta \tag{5.20}$$

with arbitrary inputs $|u_i|\infty \leq \zeta$ and $|\widehat{v}_d^i|\infty \leq \zeta$. By combining Eqs. (5.19) and (5.20), we get $(\partial\phi_{i,j}/\partial x_i^\top)(u_i - \widehat{v}_d^i) < -\gamma_2(\phi_{i,j})$, which implies that the collision-free constraints in Eq. (5.13) are naturally satisfied when $|x_{i,j}| > R$. Thus, the proof is completed. □

Remark 5.7 The elimination of non-neighboring collision-free constraints in Lemma 5.1 not only avoids the complex analysis of discontinuous situations when robots just enter the sensing radius of another robot (i.e., $|x_{i,j}| = R$), but also incorporates local sensing-only constraints. This approach can reduce communication costs and is thus practical and economical, as noted in [21].

Based on Lemma 5.1, we can concentrate on the collision-free constraints in (5.15) where $|x_{i,j}| \leq R$. However, singular cases of inter-robot collisions and robot-target overlapping (i.e., $|x_{i,j}(t)| < r$ or $|x_{i,d}(t)| = 0$, for some $i \in \mathcal{V}$ and $j \in \mathcal{N}_i$) may still occur, making the constraint-based optimization in (5.15) infeasible. Therefore, we will first address undesired collisions and overlapping (i.e., $|x_{i,j}(t)| \geq r$ and $|x_{i,d}(t)| > 0$, for all $i \neq j \in \mathcal{V}$ and $t > 0$) and then demonstrate convex-hull fencing and flexible-ordering formation in Lemmas 5.2–5.4, respectively.

Assumption 5.5 The initial values of inter-robot and robot-target distances are assumed to satisfy $|x_{i,j}(0)| \geq r$ and $|x_{i,d}(0)| > 0$ for all $i \neq j \in \mathcal{V}$.

Assumption 5.5 is essential to avoid undesired scenarios; without it, the constraint-based optimization (5.15) would lack a feasible solution from the start.

Assumption 5.6 The optimal inputs u_i^* in the constraint-based optimization (5.15) are assumed to be locally Lipschitz continuous concerning their arguments within the target set $\mathcal{T}_{i,j}$ defined in Eq. (5.12).

Assumption 5.6 is crucial for proving inter-robot collision avoidance via forward invariance in Lemma 5.2, as discussed in [22]. This assumption is necessary because the local Lipschitz continuity of u_i^* in (5.15) is not guaranteed with the additional input constraints (5.15d), which differs from the traditional QP-based problems (refer to the comment below Eq. (42) in [22]).

Lemma 5.2 *Given Assumptions 5.5 and 5.6, all robots in $\mathcal{V}$ controlled by the long-term target fencing strategy in (5.15) ensure avoidance of inter-robot collisions and prevent robot-target overlap, i.e., $|x_{i,j}(t)| \geq r$ and $|x_{i,d}(t)| > 0$ for all $i \neq j \in \mathcal{V}$ and $t > 0$.*

Proof Let $m_i - 1 \in \mathbb{Z}^+$ represent the number of neighbors in the set $\mathcal{N}_i$ for robot i, and we first rewrite the variables and functions in (5.15) as follows:

$$\begin{aligned} \delta_i &= \big[\delta_{i,0}, \overbrace{0, \ldots, 0}^{m_i - 1}\big]^{\mathrm{T}} \in \mathbb{R}^{m_i}, \; \Phi_i = \big[\phi_{i,0}, \overbrace{\ldots, \phi_{i,j}}^{m_i - 1}\big]^{\mathrm{T}} \in \mathbb{R}^{m_i}, \\ \gamma(\Phi_i) &= \big[\gamma_1(\phi_{i,0}), \overbrace{\ldots, \gamma_2(\phi_{i,j})}^{m_i - 1}\big]^{\mathrm{T}} \in \mathbb{R}^{m_i}, \end{aligned} \tag{5.21}$$

which implies that the decentralized constraint-based optimization problem (5.15) can thus be reformulated

$$\min_{u_i, \delta_i} \|u_i - \widehat{v}_d^i\|^2 + l\|\delta_i\|^2 \;\; \text{s.t.} \; g_i \leq \mathbf{0}_{m_i}, \, \mu_i \leq 0, \tag{5.22}$$

where $\mathbf{0}_{m_i} = [0, \ldots, 0]^{\mathsf{T}} \in \mathbb{R}^{m_i}$, and g_i, μ_i are given below

$$g_i := \left[\frac{\partial \Phi_i}{\partial x_i^{\mathsf{T}}}(u_i - \widehat{v}_d^i) + \gamma(\Phi_i) - \delta_i \right], \ \mu_i := \|u_i\|_\infty - \zeta$$

with ζ being the input limit of robots in (5.5). Firstly, since $\phi_{i,j} = r - \|x_{i,j}(0)\| \leq 0, \forall i \neq j \in \mathcal{V}$ in Assumption 5.5, it follows from Eq. (5.12) that $x_i(0) \in \mathcal{T}_{i,j}$. Using Assumption 5.6, it follows from the forward-invariance property in Eq. (5.4) that all the states $x_i(t)$ will stay in the set $\mathcal{T}_{i,j}$ all along, i.e., $x_i(t) \in \mathcal{T}_{i,j}, \forall i \in \mathcal{V}, t > 0$, which implies that $\|x_{i,j}(t)\| \geq r, \forall i \neq j, t > 0$. The inter-robot collision avoidance is thus proved.

Next, we aim to prevent robot-target overlapping, i.e., ensuring $|x_{i,d}(t)| > 0$ for all $t > 0$. Let the corresponding Lagrangian function for robot i in Eq. (5.22) be expressed as $\mathcal{L}(u_i, \delta_i, \lambda_i, \varsigma_i) = \|u_i - \widehat{v}_d^i\|^2 + \|\delta_i\|^2 + \lambda_i^{\mathsf{T}} g_i + \varsigma_i \mu_i$. with $\lambda_i = [\lambda_{i,1}, \ldots, \lambda_{i,m_i}]^{\mathsf{T}} \in \mathbb{R}^{m_i}$ and $\varsigma_i \in \mathbb{R}$. Using the KKT conditions [32], we take the partial derivative of $\mathcal{L}(u_i, \delta_i, \lambda_i, \varsigma_i)$ along the vector $u_i^{\mathsf{T}}, \delta_i^{\mathsf{T}}$,

$$\begin{aligned} &\begin{bmatrix} 2(u_i - \widehat{v}_d^i) \\ 2l\delta_i \end{bmatrix} + \begin{bmatrix} \frac{\partial \Phi_i^{\mathsf{T}}}{\partial x_i}\lambda_i + \varsigma_i \mathrm{sgn}(\|u_i\|_\infty) \\ -\lambda_i \end{bmatrix} = \mathbf{0}_{n+m_i}, \\ &\lambda_i^{\mathsf{T}} g_i = 0, \ \varsigma_i \mu_i = 0, \ \lambda_{i,k} \geq 0, \forall k \in \mathbb{Z}_1^{m_i}, \end{aligned} \tag{5.23}$$

which implies that

$$u_i = -l \frac{\partial \Phi_i^{\mathsf{T}}}{\partial x_i} \delta_i + \widehat{v}_d^i + \frac{\varsigma_i \mathrm{sgn}(\|u_i\|_\infty)}{2}, \ \lambda_i = 2l\delta_i. \tag{5.24}$$

From $\lambda_i^{\mathsf{T}} g_i = 0$, $\varsigma_i \mu_i = 0$ in Eq. (5.23), one has that $\lambda_i = \mathbf{0}_{m_i}$ or $g_i = \mathbf{0}_{m_i}$, and $\varsigma_i = 0$ or $\mu_i = 0$.

From the fact that the constraints in (5.22) represent target-approaching and collision-free subtasks that are eventually activated during the process, the condition $\lambda_i = \mathbf{0}m_i$ can be ruled out after a certain time $T_1 > 0$. This implies that $g_i(t) = \mathbf{0}m_i$ when $t > T_1$. Additionally, the control-input constraint is activated (i.e., $\mu_i = 0$) when the robots are far from the target, implying that there exists a time $T_2 > 0$ such that $\varsigma_i = 0$ holds when $t > T_2$. Letting $T_3 := \max\{T_1, T_2\}$, it follows that $\|x_{i,j}(t)\| \geq r$ and $|x_{i,d}(t)| > 0$ for all $i \neq j \in \mathcal{V}$ and $t \in [0, T_3]$ because u_i is bounded and the constraints g_i are not activated. Next, we will prove the absence of undesirable overlapping for $t > T_3$. Given $g_i = \mathbf{0}_{m_i}$ and $\varsigma_i = 0$, it follows from (5.24) that

$$-l \frac{\partial \Phi_i}{\partial x_i^{\mathsf{T}}} \frac{\partial \Phi_i^{\mathsf{T}}}{\partial x_i} \delta_i + \gamma(\Phi_i) - \delta_i = \mathbf{0}_{m_i}. \tag{5.25}$$

Combining with $\partial \Phi_i/\partial x_i^\top = (\partial \Phi_i^\top/\partial x_i)^\top \in \mathbb{R}^{m\times n}$ and (5.25) yields

$$\delta_i = \Xi_i \gamma(\Phi_i) \tag{5.26}$$

with

$$\Xi_i = \left(I_m + l\frac{\partial \Phi_i}{\partial x_i^\top}\left(\frac{\partial \Phi_i}{\partial x_i^\top}\right)^\top\right)^{-1} \in \mathbb{R}^{m\times m}. \tag{5.27}$$

Substituting $\varsigma_i = 0$ and Eq. (5.26) into Eq. (5.24) yields the optimal input

$$u_i^* = -l\frac{\partial \Phi_i^\top}{\partial x_i}\Xi_i \gamma(\Phi_i) + \widehat{v}_d^i. \tag{5.28}$$

According to the implicit function $\gamma(\Phi_i)$, $i \in \mathcal{V}$ in Eq. (5.21), a Lyapunov function candidate is picked below

$$V = \sum_{i=1}^{N} \gamma_1(\phi_{i,0})^2 + \frac{1}{2}\sum_{i=1}^{N}\sum_{j\in\mathcal{N}_i} \gamma_2(\phi_{i,j})^2, \tag{5.29}$$

whose derivative is

$$\dot{V} = \sum_{i=1}^{N} \gamma_1(\phi_{i,0})\frac{\partial \gamma_1(\phi_{i,0})}{\partial x_{i,d}^\top}(\dot{x}_i - \dot{x}_d) + \frac{1}{2}\sum_{i=1}^{N}\sum_{j\in\mathcal{N}_i} \gamma_2(\phi_{i,j})\frac{\partial \gamma_2(\phi_{i,j})}{\partial x_{i,j}^\top}(\dot{x}_i - \dot{x}_j). \tag{5.30}$$

Recalling the definition of $\phi_{i,j}$ in Eq. (5.13), one has that

$$\sum_{i=1}^{N}\sum_{j\in\mathcal{N}_i} \gamma_2(\phi_{i,j})\frac{\partial \gamma_2(\phi_{i,j})}{\partial x_{i,j}^\top}(\dot{x}_i - \dot{x}_j) = 2\sum_{i=1}^{N}\sum_{j\in\mathcal{N}_i} \gamma_2(\phi_{i,j})\frac{\partial \gamma_2(\phi_{i,j})}{\partial x_{i,j}^\top}\dot{x}_i. \tag{5.31}$$

Moreover, we have $\sum_{i=1}^{N}\sum_{j\in\mathcal{N}_i} \gamma_2(\phi_{i,j})(\partial \gamma_2(\phi_{i,j})/\partial x_{i,j}^\top)\ v_d = 0$. Substituting Eqs. (5.5), (5.7), (5.31) into Eq. (5.30) yields

$$\dot{V} = \sum_{i=1}^{N}\left\{\left(\gamma_1(\phi_{i,0})\frac{\partial \gamma_1(\phi_{i,0})}{\partial x_{i,d}^\top} + \sum_{j\in\mathcal{N}_i} \gamma_2(\phi_{i,j})\frac{\partial \gamma_2(\phi_{i,j})}{\partial x_{i,j}^\top}\right)(u_i - v_d)\right\}. \tag{5.32}$$

Meanwhile, from the definition of $\phi_{i,0}$, $\phi_{i,j}$ in Eqs. (5.9) and (5.12), one has that

$$\frac{\partial \gamma_1(\phi_{i,0})}{\partial x_{i,d}^\top} = \frac{\partial \gamma_1(\phi_{i,0})}{\partial \phi_{i,0}}\frac{\partial \phi_{i,0}}{\partial x_i^\top},\ \frac{\partial \gamma_2(\phi_{i,j})}{\partial x_{i,j}^\top} = \frac{\partial \gamma_2(\phi_{i,j})}{\partial \phi_{i,j}}\frac{\partial \phi_{i,j}}{\partial x_i^\top}. \tag{5.33}$$

Recalling $\gamma_1(\cdot)$, $\gamma_2(\cdot)$ in Eqs. (5.11) and (5.14) are locally Lpischiz, there exists a constant $B > 0$ satisfying

$$\frac{\partial \gamma_1(\phi_{i,0})}{\partial \phi_{i,0}^{\mathsf{T}}} \leq B, \frac{\partial \gamma_2(\phi_{i,j})}{\partial \phi_{i,j}^{\mathsf{T}}} \leq B, \forall i \in \mathcal{V}, j \in \mathcal{N}_i, \tag{5.34}$$

it then follows from Eqs. (5.21), (5.28), (5.33), and (5.34) that $\dot{V}$ in Eq. (5.32) rewrites

$$\dot{V} \leq -B \sum_{i=1}^{N} \left(\frac{\partial \Phi_i^{\mathsf{T}}}{\partial x_i} \gamma(\Phi_i) \right)^{\mathsf{T}} \left(l \frac{\partial \Phi_i^{\mathsf{T}}}{\partial x_i} \Xi_i \gamma(\Phi_i) - \widetilde{v}_d^i \right) \tag{5.35}$$

with $\widetilde{v}_d^i := \widehat{v}_d^i - v_d$. Meanwhile, according to Woodbury matrix identity [33], one has that Ξ_i in (5.27) is rewritten as

$$\Xi_i = I_{m_i} - l \frac{\partial \Phi_i}{\partial x_i^{\mathsf{T}}} \widetilde{\Xi}_i \frac{\partial \Phi_i}{\partial x_i^{\mathsf{T}}} \tag{5.36}$$

with

$$\widetilde{\Xi}_i := \left(I_n + l \left(\frac{\partial \Phi_i}{\partial x_i^{\mathsf{T}}} \right)^{\mathsf{T}} \frac{\partial \Phi_i}{\partial x_i^{\mathsf{T}}} \right)^{-1} \in \mathbb{R}^{n \times n}. \tag{5.37}$$

By combining Eqs. (5.36) and (5.37), we obtain $(\partial \Phi_i^{\mathsf{T}}/\partial x_i)\Xi_i = \widetilde{\Xi}_i(\partial \Phi_i^{\mathsf{T}}/\partial x_i)$. Since the matrix $\widetilde{\Xi}_i$ in Eq. (5.27) is positive definite, it follows that the smallest eigenvalue $\lambda_{min}(\widetilde{\Xi}_i) > 0$, indicating that

$$\left(\frac{\partial \Phi_i^{\mathsf{T}}}{\partial x_i} \gamma(\Phi_i) \right)^{\mathsf{T}} \widetilde{\Xi}_i \left(\frac{\partial \Phi_i^{\mathsf{T}}}{\partial x_i} \gamma(\Phi_i) \right) \geq \lambda_{min}(\widetilde{\Xi}_i) \left\| \frac{\partial \Phi_i^{\mathsf{T}}}{\partial x_i} \gamma(\Phi_i) \right\|^2 \tag{5.38}$$

based on Courant–Fischer Theorem [34]. At the same time, one has

$$\left(\frac{\partial \Phi_i^{\mathsf{T}}}{\partial x_i} \gamma(\Phi_i) \right)^{\mathsf{T}} \widetilde{v}_d^i \leq \frac{1}{\kappa} \left\| \frac{\partial \Phi_i^{\mathsf{T}}}{\partial x_i} \gamma(\Phi_i) \right\|^2 + \kappa \|\widetilde{v}_d^i\|^2 \tag{5.39}$$

with an arbitrarily selected constant $\kappa > 0$ satisfying $1/\kappa < l\lambda_{min}(\widetilde{\Xi}_i)$, which then follows from Eqs. (5.35), (5.38), (5.39) that $\dot{V} \leq -\alpha + \beta$ where

$$\alpha := B \sum_{i \in \mathcal{V}} \left(l\lambda_{min}(\widetilde{\Xi}_i) - \frac{1}{\kappa} \right) \left\| \frac{\partial \Phi_i^{\mathsf{T}}}{\partial x_i} \gamma(\Phi_i) \right\|^2 \geq 0, \;\; \beta := B \sum_{i \in \mathcal{V}} \kappa \|\widetilde{v}_d^i\|^2 \geq 0. \tag{5.40}$$

By applying the comparison theorems from [35] and integrating $\dot{V}$ from Eq. (5.35) with respect to time t, we derive

$$V(t) \leq V(T_3) - \int_{T_3}^{t} \alpha(s)ds + \int_{T_3}^{t} \beta(s)ds,\ t > T_3. \tag{5.41}$$

Leveraging Assumption 5.5, we can see from Eqs. (5.11), (5.14), and (5.29) that the initial value of $V(T_3)$ is constrained. According to Eq. (5.40), we have $-\int_{T_3}^{t} \alpha(s)ds \leq 0$. Given that $\widetilde{v}_d^i(t)$ approaches $\mathbf{0}n$ exponentially as $t \to \infty$, as noted in Remark 5.4, Eq. (5.40) indicates that $\lim_{t\to\infty} \beta(t) = 0$ exponentially. This implies that $\int_{T_3}^{t} \beta(s)ds$ is bounded above, leading to the conclusion that $V(t)$ is also bounded.

The function $V(t)$ only becomes unbounded if $|x_{i,j}(t)| = 0$ for all $j \in \mathcal{N}_i$ or if $|x_{i,d}(t)| \leq 0$. However, this contradicts the result from Eq. (5.41) which shows that $V(t)$ is bounded. Consequently, we can conclude that $|x_{i,j}(t)| \geq r$ and $|x_{i,d}(t)| > 0$ for all $i \neq j \in \mathcal{V}$ and for all $t > 0$. Thus, the proof is completed. □

Remark 5.8 When a global reference frame is not available for the robots, it is still feasible to apply the velocity estimation $\widehat{v}_d^i$ as discussed in Remark 5.4 and the long-term target fencing problem in Eq. (5.15). Specifically, suppose robot i operates in its own local frame, which is obtained by rotating the global frame with the matrix $\mathcal{R}_i \in \mathbb{R}^{n\times n}$ and translating it by the vector $p_i \in \mathbb{R}^n$. In this local frame, the states are $(\widehat{x}_d^i)^{[i]} = \mathcal{R}_i\widehat{x}_d^i + p_i$, $(\widehat{v}_d^i)^{[i]} = \mathcal{R}_i\widehat{v}_d^i$, $x_d^{[i]} = \mathcal{R}_i x_d + p_i$, $v_d^{[i]} = \mathcal{R}_i v_d$, $x_l^{[i]} = \mathcal{R}_i x_l + p_i$, $l \in \mathbb{Z}_1^N$, $u_i^{[i]} = \mathcal{R}_i u_i$, where the superscript $[i]$ denotes the ith local frame. Then, the example of velocity estimator in Eq. (5.8) becomes

$$\begin{aligned}(\dot{\widehat{x}}_d^i)^{[i]} &= -\chi_1((\widehat{x}_d^i)^{[1]} - x_d^{[1]}) + (\widehat{v}_d^i)^{[i]},\\ (\dot{\widehat{v}}_d^i)^{[i]} &= -\chi_1\chi_2((\widehat{x}_d^i)^{[i]} - x_d^{[i]}),\end{aligned}$$

which also achieves that $\lim_{t\to\infty}\{(\widehat{v}_d^i)^{[i]}(t) - v_d^{[i]}(t)\} = \mathbf{0}_n$, exponentially. It follow from Eqs. (5.9), (5.11), (5.12), and (5.14) that $\|u_i - v_d^i\| = \|u_i^{[i]} - (\widehat{v}_d^i)^{[i]}\|$, $\phi_{i,0} = \phi_{i,0}^{[i]}$, $\phi_{i,j} = \phi_{i,j}^{[i]}$, which implies that the proposed framework in Eq. (5.15) can be implemented in the ith local frame, i.e.,

$$\begin{aligned}&\min_{u_i^{[i]},\delta_{i,0}} \left\{\|u_i^{[i]} - (\widehat{v}_d^i)^{[i]}\|^2 + l\delta_{i,0}^2\right\}\\ &\text{s.t. } \frac{\partial\phi_{i,0}^{[i]}}{\partial(x_i^{[i]})^\top}(u_i^{[i]} - (\widehat{v}_d^i)^{[i]}) + \gamma_1(\phi_{i,0}^{[i]}) - \delta_{i,0} \leq 0,\\ &\quad \frac{\partial\phi_{i,j}^{[i]}}{\partial(x_i^{[i]})^\top}(u_i^{[i]} - (\widehat{v}_d^i)^{[i]}) + \gamma_2(\phi_{i,j}^{[i]}) \leq 0,\ j \in \mathcal{N}_i,\\ &\quad \|u_i^{[i]}\|_\infty \leq \zeta, \forall i \in \mathcal{V}.\end{aligned}$$

Lemma 5.3 *Given Assumption 5.1, the convex-hull fencing is achieved by all robots $\mathcal{V}$ under the long-term target fencing problem in Eq. (5.15). Specifically, it holds that* $\lim_{t\to\infty}\left\{\sum_{i=1}^N x_i(t)/N - x_d(t)\right\} = \mathbf{0}_n$.

Proof Given that $V(t)$ is upper bounded as shown in Lemma 5.2, it follows from Eqs. (5.21) and (5.29) that both $\gamma(\Phi_i)$ and Φ_i are also bounded. Additionally, by examining the definition of α in Eq. (5.40), $V(t)$ in Eq. (5.41) can be expressed below

$$\int_{T_3}^{t} \alpha(s)ds \leq V(T_3) - V(t) + \int_{T_3}^{t} \beta(s)ds, \tag{5.43}$$

which implies that $\int_{T_3}^{t} \alpha(s)ds$ is upper bounded and reaches a finite value as $t \to \infty$. Since $\dot{\alpha}$ depends on $\gamma(\Phi_i)$, $\dot{\gamma}(\Phi_i)$, $\dot{\Phi}_i$, and $\ddot{\Phi}_i$, all of which are bounded, α is uniformly continuous. Consequently, by applying Barbalat's lemma [35], it follows that $\lim_{t\to\infty} \alpha(t) = 0$. According to the definition of α in Eq. (5.40), it can be deduced that

$$\lim_{t\to\infty} \frac{\partial \Phi_i^{\top}(t)}{\partial x_i(t)} \gamma(\Phi_i(t)) = \mathbf{0}_n, \forall i \in \mathcal{V}. \tag{5.44}$$

It follows from Φ_i, $\gamma(\Phi_i)$ in Eq. (5.21) that

$$\lim_{t\to\infty} \left\{ \frac{\partial \phi_{i,0}(t)}{\partial x_i^{\top}(t)} \gamma_1(\phi_{i,0}(t)) + \sum_{j\in\mathcal{N}_i} \frac{\partial \phi_{i,j}(t)}{\partial x_i^{\top}(t)} \gamma_2(\phi_{i,j}(t)) \right\} = \mathbf{0}_n. \tag{5.45}$$

With $\phi_{i,0} = -h_{i.0}, \phi_{i,j} = -h_{i,j}$, $j \in \mathcal{N}_i$ in Eqs. (5.9), (5.12), one has that

$$\frac{\partial \phi_{i,0}(t)}{\partial x_i^{\top}(t)} = \frac{x_{i,d}^{\top}}{\|x_{i,d}\|}, \quad \frac{\partial \phi_{i,j}(t)}{\partial x_i^{\top}(t)} = \frac{-x_{i,j}}{\|x_{i,j}\|}, \; j \in \mathcal{N}_i. \tag{5.46}$$

Combining with Eqs. (5.45) and (5.46) yields

$$\lim_{t\to\infty} \left\{ \frac{x_{i,d}(t)}{\|x_{i,d}(t)\|} \gamma_1\big(\|x_{i,d}(t)\|\big) - \sum_{j\in\mathcal{N}_i} \frac{x_{i,j}(t)}{\|x_{i,j}(t)\|} \gamma_2\big(r - \|x_{i,j}(t)\|\big) \right\} = \mathbf{0}_n, i \in \mathcal{V}. \tag{5.47}$$

Note that for arbitrary two neighboring robots $i \neq j \in \mathcal{V}$, one has $\gamma_2\big(r - \|x_{i,j}\|\big)x_{i,j}/\|x_{i,j}\| = -\gamma_2\big(r - \|x_{j,i}\|\big)x_{j,i}/\|x_{j,i}\|$, which implies that $\sum_{i=1}^{N}\sum_{j\in\mathcal{N}_i} \gamma_2\big(r - \|x_{i,j}(t)\|\big)x_{i,j}(t) / \|x_{i,j}(t)\| = \mathbf{0}_n, \forall t > 0$. Then, it follows from Eqs. (5.45) and (5.47) that $\lim_{t\to\infty} \sum_{i=1}^{N} \{\gamma_1\big(\|x_{i,d}(t)\|\big)x_{i,d}/\|x_{i,d}\|\} = \mathbf{0}_n$.

Given that $\gamma_1(|x_{i,d}(t)|)$ in Eq. (5.11) is an odd function, it follows that

$$\lim_{t\to\infty} \sum_{i=1}^{N} x_{i,d}(t) = \mathbf{0}_n.$$

Let the target-fencing errors be $e := 1/N \sum_{i=1}^{N} x_i - x_d$, one has that

$$\lim_{t \to \infty} e(t) = \lim_{t \to \infty} \frac{1}{N} \sum_{i=1}^{N} \left\{ x_i(t) - x_d(t) \right\} = \mathbf{0}_n \tag{5.48}$$

with $1/N \sum_{i=1}^{N} x_d = x_d$. The proof of convex-hull fencing is thus completed. □

Lemma 5.4 *Given Assumption 5.5, the long-term target fencing described in (5.15) ensures that all robots $\mathcal{V}$ achieve a flexible-ordering formation, which means that*

$$\begin{aligned} &(a) \lim_{t \to \infty} \dot{x}_{s[i],s[i+1]}(t) = \mathbf{0}_n, \\ &(b)\ 0 < \lim_{t \to \infty} \|x_{s[i],s[i+1]}(t)\| < R, \\ &\forall i \in \mathbb{Z}_1^N \ (\textit{If } i = N, \textit{then } s[i+1] = s[1]), \end{aligned} \tag{5.49}$$

where $x_{s[i],s[i+1]} := x_{s[i]} - x_{s[i+1]}$.

Proof According to (5.28) and (5.44) in Lemmas 5.2 and 5.3 respectively, one has that $\lim_{t \to \infty} u_i^*(t) = \widehat{v}_d^i, \forall i \in \mathcal{V}$. Combining with the condition of $\lim_{t \to \infty} v_d^i(t) = v_d(t)$ in Remark 5.4, it can be deduced that the optimal input u_i^* of robot i satisfy $\lim_{t \to \infty} u_i^*(t) = \lim_{t \to \infty} u_j^*(t) = v_d(t), \forall i \neq j \in \mathcal{V}$, which follows from Eq. (5.5) that $\lim_{t \to \infty} \{\dot{x}_i(t) - \dot{x}_j(t)\} = \mathbf{0}_n$, (i.e., the relative position of x_i against any other x_j, $j \neq i$ converges to be time-invariant and a rigid pattern with a spatial sequence $s[1], s[2], \cdots, s[N]$ is formed). The condition (a) in (5.49) is thus verified.

The condition (b) in Eq. (5.49) will be established. According to Lemma 5.2, it follows that $|x_{i,j}(t)| \geq r$ and $|x_{i,d}(t)| > 0$ for all $i \neq j \in \mathcal{V}$ and $t > 0$. This implies that

$$\lim_{t \to \infty} |x_{s[i],s[i+1]}(t)| \geq r, \forall i \in \mathbb{Z}_1^N (\text{if } i = N, \text{ then } s[i+1] = s[1]). \tag{5.50}$$

Thus, the left-side inequality in Eq. (5.49) is satisfied. We will now prove the right-side inequality of condition (b) in Eq. (5.49) by contradiction.

Intuitively, from Lemma 5.3, we obtain that the target x_d is asymptotically fenced in the centroid of the convex hull $\sum_{i=1}^{N} x_i/N$ with a spatial sequence $s[1], s[2], \cdots, s[N]$. Without loss of generality, we assume that there exists at least one pair of adjacent robots labeled $s[l], s[l+1], \forall l \in \mathbb{Z}_1^{N-1}$ such that $\|x_{s[l],s[l+1]}\| > R$, and then separate the robots $\mathcal{V}$ into two subgroups $\mathcal{V}_1 := \{s[l+1], s[l+2], \cdots, s[k]\}$ and $\mathcal{V}_2 := \{s[l-1], s[l-2], \cdots, s[k+1]\}$ along the dashed line (i.e., a dashed plane in 3D) formed by $x_{s[l]}$ and x_d, as shown in Fig. 5.4. Then, the contradiction is analyzed in the following two cases.

Case 1: {Robot $s[k+1]$ is not on the dashed line, see Fig. 5.4a}. For $\|x_{s[l],s[l+1]}\| > R$, one has that $\|x_{s[l],j}\| > R$, $j \in \mathcal{V}_1$, which implies that robot $s[l]$ may only

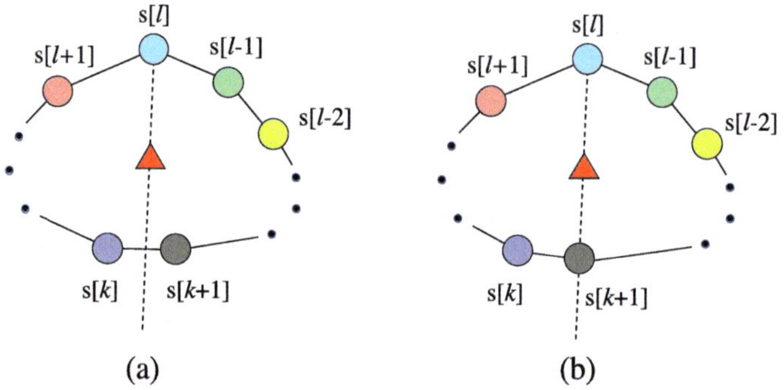

Fig. 5.4 2D illustration of two-subgroup robots along the dashed line formed by $x_s[l]$ and x_d

have neighbors fulfilling $\|x_{s[l],o}\| \leq R, o \in \mathcal{V}_2$. Moreover, the vectors $x_{s[l],d}$ and $x_{s[l],o}$, $o \in \mathcal{V}_2$, are not in same or opposite direction, i.e.,

$$-1 < \lim_{t\to\infty} \frac{x_{s[l],d}(t) \cdot x_{s[l],o}(t)}{\|x_{s[l],d}(t)\| \|x_{s[l],o}(t)\|} < 1, o \in \mathcal{V}_2, \tag{5.51}$$

Let ϑ_o, $o \in \mathcal{V}_2$, be the repulsion term between robots $s[l]$ and $s[o]$, $o \in \mathcal{V}_2$ below,

$$\vartheta_o = \frac{x_{s[l],o}}{\|x_{s[l],o}\|}\gamma_2\big(r - \|x_{s[l],o}(t)\|\big), o \in \mathcal{V}_2, \tag{5.52}$$

it follows from Eq. (5.51) that there only exist left-side repulsion terms ϑ_o^L, $o \in \mathcal{V}_2$, perpendicular to $x_{s[l],d}$ below,

$$\vartheta_o^L = \vartheta_o - \frac{x_{s[l],d}}{\|x_{s[l],d}\|} \cdot \vartheta_o. \tag{5.53}$$

However, ϑ_o^L cannot be eliminated by the robot-target attraction term $(x_{s[l],d}/\|x_{s[l],d}\|)\,\gamma_1\big(\|x_{s[l],d}\|\big)$. Then, one has

$$\lim_{t\to\infty}\left\{\frac{x_{s[l],d}(t)}{\|x_{s[l],d}(t)\|}\gamma_1\big(\|x_{s[l],d}(t)\|\big) - \sum_{o\in\mathcal{V}_2}\vartheta_o\right\} \neq \mathbf{0}_n.$$

Then, it contradicts the condition in Eq. (5.47).

Case 2: {Robot $s[k+1]$ is on the dashed line, see Fig. 5.4b}. Since robot $s[k+1]$ is on the dashed line, one has that $x_{s[l],s[k+1]}$ and $x_{s[l],d}$ are in the opposite direction, which follows from Eqs. (5.52) and (5.53) that $\vartheta_{s[k+1]}^L = \vartheta_{s[k+1]} - (x_{s[l],d}/\|x_{s[l],d}\|) \cdot \vartheta_o = \mathbf{0}_n$. Then, the repulsion term $\vartheta_{s[k+1]}$ cannot eliminate the other left-side repulsion term ϑ_o^L in Fig. 5.4b as well. The following contradiction analysis is the same as Case 1, which is omitted. Finally, we obtain that $\lim_{t\to\infty} \|x_{s[i],s[i+1]}(t)\| \leq R, \forall i \in \mathbb{Z}_1^N$ (If $i = N$, then $s[i+1] = s[1]$). The proof is thus completed. □

Remark 5.9 In the optimization problem (5.15), the time-varying neighbor set $\mathcal{N}_i(t)$ makes the collision-free constraint (5.15c) time-dependent, implying that the repulsion forces between neighboring robots, dictated by these constraints, also vary over time. At the same time, when combined with the attraction forces toward the target from constraint (5.15b), the system achieves a dynamic equilibrium similar to physical forces. This results in a solution to the optimization problem (5.15) that is flexible, thereby achieving *ordering-flexible* fencing.

Theorem 5.1 *Under Assumptions 5.1–5.6, a multi-robot system in (5.5) governed by the constraint-based optimization problem (5.15) achieves the long-term ordering-flexible target fencing.*

Proof We draw the conclusion from Lemmas 5.1–5.4 directly. □

Remark 5.10 Although the CBF constraints in Eqs. (5.15b)–(5.15c) are established only based on position, where the implicit regulation of the robot's orientation is still possible. One common approach is to convert the high-level desired velocity u_i^* in Eq. (5.15) into the desired orientation, such as the desired yaw angle $\theta_i^* = \arctan 2(u_{i,2}^*, u_{i,1}^*)$ in the 2D plane. This desired orientation can then be tracked using an additional low-level tracking module [2]. With all robots maintaining a rigid pattern around the target, i.e., $u_i^* = v_d, i \in \mathcal{V}$, their orientations will converge to be the same. For robots with simple unicycle dynamics, Sect. 5.4.1 will use near-identity diffeomorphism (NID) [36] to regulate their orientation. Moreover, this chapter primarily focuses on homogeneous robots as a preliminary exploration of ordering-flexible target fencing. Although two types of AMRs are used in the experiments of Sect. 5.4.1, their tracking capabilities are nearly identical, thereby they do not affect the fencing performance. However, extending the proposed LTTE algorithm (5.15) to heterogeneous multi-robot systems remains challenging due to the diverse capabilities of different robots, and this will be addressed in future research.

Remark 5.11 The proposed framework in Eq. (5.15) offers three main benefits: (i) **Feasibility**: The introduction of slack variables in the constraints ensures that a feasible solution always exists for multi-robot long-term target fencing. For further details, refer to Remark 5.5. (ii) **Forward Invariance**: By incorporating different target-fencing subtasks as constraints rather than cost functions, Eq. (5.15) retains the forward-invariance property of Eq. (5.3) as defined in Definition 5.1. This property ensures rigorous collision avoidance, as demonstrated in Lemma 5.2. (iii) **Robustness**: The use of a time-varying neighbor set $\mathcal{N}_i(t)$ allows the system to adapt even if some robots suddenly break down. The remaining robots governed by Eq. (5.15) can still achieve the *ordering-flexible* target fencing, as illustrated in Fig. 5.14.

Remark 5.12 The potential undesired equilibria commonly observed in QP-CLF-CBF and QP-CBF problems refer to scenarios where robots converge to the boundary of the safe set rather than the minimum of the Lyapunov function [37]. Specifically, these undesired equilibria in Eq. (5.15) are characterized by $\{\|x_{i,j}\| = r, \|x_{i,d}\| = \delta_{i,0}/\eta_1 > 0$ with $\delta_{i,0} > 0$ remaining constant for $i \in \mathcal{V}$ and $j \in \mathcal{N}_i$, and where the constraints (5.15b) and (5.15c) are active. Despite these undesired equilibria, they still satisfy the conditions for fencing formation and do not affect the performance of the target-fencing tasks. Unlike traditional QP-CBF approaches [15, 16, 37] which focus on approaching the target point (i.e., $|x_{i,d}| = 0$ in this article), the proposed optimization problem (5.15) aims to achieve a fencing formation (i.e., (i.e., $\sum_{i\in\mathcal{V}} x_i/N = \mathbf{0}_n$, $\|x_{i,j}\| \geq r$, $\|x_{i,d}\| > 0$ as defined in Definition 5.2) by balancing the target-approaching and collision-free constraints (5.15b)–(5.15c). Thus, the *ordering-flexible* fencing formation can still be realized even with the undesired equilibria of $\{\|x_{i,j}\| = r, \|x_{i,d}\| = \delta_{i,0}/\eta_1 > 0, \delta_{i,0} > 0$ keeps invariant, $\forall i \in \mathcal{V}, j \in \mathcal{N}_i$. It is illustrated in Fig. 5.5, where both triangle- and square-pattern fencing formations are achieved under these undesired equilibria.

Remark 5.13 Based on the constraint-driven optimization setup in (5.15), we can further tackle the static obstacle avoidance by seamlessly accommodating the corresponding robot-obstacle subtask $\mathcal{T}^o_{i,k}$, $k \in \mathcal{O}$, to be $\phi^o_{i,k} = \|x^o_{i,k}\| - r^o_k$ with $x^o_{i,k} := x_i - x^o_k$, the position x^o_k and the radius r^o_k to cover the kth circular obstacle in the obstacle set $\mathcal{O}$. More precisely, the condition of $\phi^o_{i,k}$ is encoded to be

$$\frac{\partial \phi^o_{i,k}}{\partial x_i^\top}(u_i - \widehat{v}^i_d) \leq -\gamma_2(\phi^o_{i,k}), i \in \mathcal{V}, k \in \mathcal{O}, \tag{5.54}$$

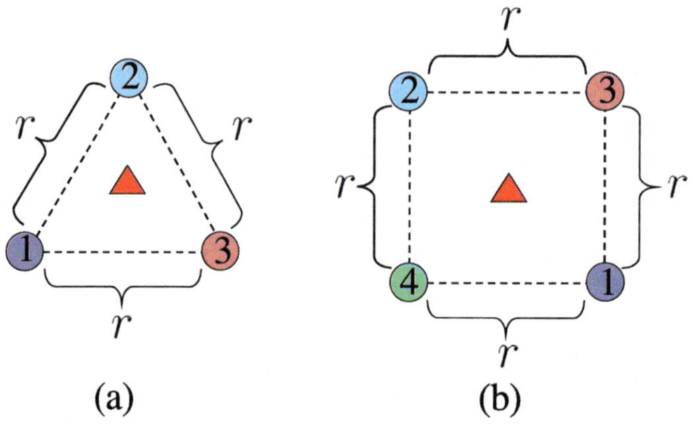

Fig. 5.5 **a** The triangle-pattern fencing is achieved under the undesired equilibria: $\{\|x_{1,2}\| = \|x_{2,3}\| = \|x_{1,3}\| = r, \|x_{i,d}\| > 0, i = 1, 2, 3\}$. **b** The square-pattern fencing is achieved under the undesired equilibria: $\{\|x_{1,4}\| = \|x_{1,3}\| = \|x_{2,4}\| = \|x_{2,3}\| = r, \|x_{i,d}\| > 0, i = 1, 2, 3, 4\}$. (All the symbols have the same meaning as in Fig. 5.3)

which can be added into the constraints of the optimization problem (5.15). For the feasibility of Eq. (5.15) with the additional obstacle-avoidance constraints (5.54), based on the initial condition of $\phi^o_{i,k}(0) > 0$ and forward-invariance property in Eq. (5.4), there also exists a feasible solution $\{u_i = \widehat{v}^i_d, \delta_{i,0} > 0$ is sufficiently large$\}$, such that all the constraints (5.15b)–(5.15d), and (5.54) are satisfied, which is similar to Remark 5.5. For the undesired local equilibria of Eq. (5.15) with the additional obstacle-avoidance constraints (5.54), robots cannot stay at the undesired equilibrium points and will finally leave them. Specifically, after adding the constraints (5.54), it follows from [37] that the undesired equilibrium points become

$$\begin{aligned}\{&\|x^o_{i,k}\| = r^o_k, \|x_{i,j}\| = r, \|x_{i,d}\| = \delta_{i,0}/\eta_1 \neq 0,\\ &\text{and } \delta_{i,0} > 0 \text{ keeps invariant}, i \in \mathcal{V}, j \in \mathcal{N}_i\}\end{aligned} \quad (5.55)$$

with the constraints (5.15b), (5.15c) and (5.54) being active. We assume that robots are already at the undesired equilibria when $t = T_3 > 0$. (i) Since the target x_d is moving with the velocity v_d, one has that x_i will moving to satisfy $\|x_{i,d}\| = \delta_{i,0}/\eta_1$ in Eq. (5.55). (ii) However, due to the obstacles being static, it follows from $\|x^o_{i,k}\| = x_i - x^o_k = r^o_k$ in Eq. (5.55) that x_i cannot deviate from the circle where the obstacle x^o_k is located, which contradicts (i). Hence, the undesired equilibrium points will be excluded after some time $t > T_3$. For more complex scenarios of moving obstacles, the feasibility and undesired equilibrium are hard to analyze and will be investigated in future works.

5.4 Algorithm Verification

In this section, we will conduct 2D experiments and 3D simulation for algorithm verification.

5.4.1 2D Experiments

In this subsection, we present 2D experiments conducted with a multi-AMR platform to assess the effectiveness of the long-term *ordering-flexible* target fencing (5.15). We begin by describing the multi-AMR platform. As depicted in Fig. 5.6a, the platform includes a 5 m × 5 m working space and three AMRs: the largest being Hunter 1.0 and the other two being Scout Mini.[2] In Fig. 5.6b, it is shown that the Hunter-1.0 AMR measures 0.98 m in length and 0.72 m in width, while the Scout Mini AMR measures 0.62 m in length and 0.54 m in width. Each AMR is equipped

[2] AMRs: https://global.agilex.ai/.

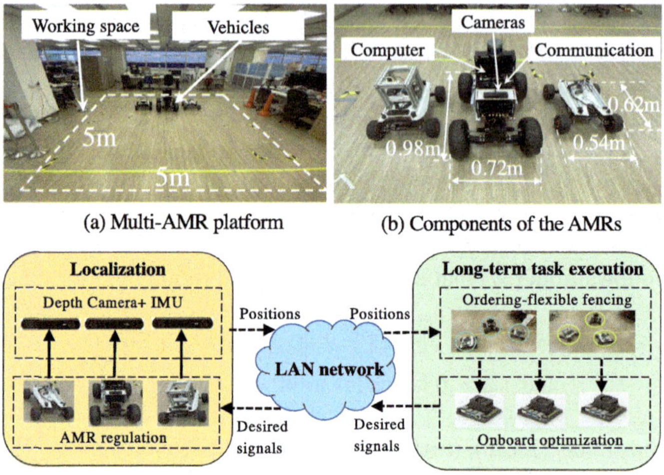

(c) The operation procedure of the long-term target fencing

Fig. 5.6 **a** The multi-AMR platform consists of three AMRs and a working space. **b** Sizes of two kinds of AMRs and detailed components. **c** The operation procedure of the target fencing consists of localization, LAN network, and long-term task execution. (The solid arrows in subfigure **c** represent the physical connection, and the dashed arrows in subfigure **c** are the virtual connection.)

with an onboard computer, the NVIDIA Jetson Xavier NX,[3] a depth stereo camera, the ZED2*i*,[4] which integrates depth detection and an inertial measurement unit (IMU), and a ROS TCP-protocol communication component. Figure 5.6c details the operational procedure, where the AMR uses the ZED2i camera for position localization via a VINS-Fusion SLAM algorithm [38], and then transmits this position data to the onboard computer via a common WiFi LAN network. The onboard computer calculates the desired signals based on its own position and those of its neighbors, which are then communicated to the AMR actuators. It is noteworthy that AGILE·X company has effectively implemented the tracking of these desired signals in the AMR regulation module.

To adapt the optimal u_i^* in Eq. (5.15) for commercial AMRs with specified speeds and rotation rates, we use the unicycle model for AMRs as described in [17]

$$\dot{x}_{i,1} = v_i \cos\theta_i,\ \dot{x}_{i,2} = v_i \sin\theta_i,\ \dot{\theta}_i = u_{\theta_i}, i \in \mathcal{V}, \tag{5.56}$$

where $x_i = [x_{i,1}, x_{i,2}]^\top \in \mathbb{R}^2$ represents the position, $\theta_i \in \mathbb{R}$ is the yaw angle in the X-Y plane, and $v_i \in \mathbb{R}$ and $u_{\theta_i} \in \mathbb{R}$ are the desired velocity and rotation rate for robot i, respectively. Using near-identity diffeomorphism (NID) [36], the desired velocities

[3] NVIDIA Jetson Xavier NX: https://www.nvidia.com/en-us/autonomous-machines/embedded-systems.

[4] ZED2*i*: https://www.stereolabs.com/zed-2i/.

$u_{i,1}^*, u_{i,2}^*$ from Eq. (5.15) can be converted into the desired commands v_i, u_{θ_i} as given in Eq. (5.56) below.

$$\begin{aligned} v_i &= u_{i,1}^* \cos\theta_i + u_{i,2}^* \sin\theta_i, \\ u_{\theta_i} &= -\frac{u_{i,1}^*}{l}\sin\theta_i + \frac{u_{i,2}^*}{l}\cos\theta_i, i \in \mathcal{V}, \end{aligned}$$

with $l \in \mathbb{R}^+$ being an arbitrary small constant. In the experiments, l is set to be $l = 0.2$ for convenience.

In the following, we address the *ordering-flexible* fencing task with three AMRs and a constant-velocity virtual target, considering the constraints of the limited working space. Based on Eq. (5.5), the input limit is set to $\zeta = 0.2$, meaning $\|u_i\|_\infty \le 0.2, i \in \mathbb{Z}_1^3$. To prevent the Scout Mini and Hunter AMRs from exceeding the working space and to accommodate their varying performance characteristics, we set $\xi = 0.2$, which is much lower than the maximum speeds of the Scout Mini (2.7 m/s) and Hunter 1.0 (4.8 m/s). This ensures that the AMRs operate at lower speeds with minimal turning rates. Additionally, setting ξ uniformly across all AMRs provides similar performance capabilities. The selection of parameters r, R, and η_2 is as follows: (i) Considering the sizes of the Scout Mini and Hunter AMRs, we choose the collision radius $r = 1.5$ m. (ii) Based on the effective range of the ZED2i camera, we set the sensing radius to $R = 4$ m. (iii) According to Assumption 5.4, η_2 is determined as $R - r = 2.5$. The gain η_1 in Eq. (5.10) is set to 2.0, and the initial value of the slack variable $\delta_{i,0}$ is 100. Due to the large size of the AMRs and the wide turning radius (1.6 m) of Hunter 1.0, which occupies significant space, the virtual target's velocity is set to a constant $v_d = [0.06, 0]^\top$ m/s. A distributed velocity estimator with an additional connected communication topology, as shown in [31], is used to track the target's velocity. Figure 5.7 presents two experimental scenarios (Cases 1 and 2) demonstrating the three AMRs collectively achieving the fencing formation around the moving target. Specifically, Fig. 5.7a and b show that starting from different initial positions (depicted as blue vehicles) and fulfilling Assumption 5.5, the AMRs successfully form a triangle-pattern target fencing (depicted as red vehicles) with distinct orderings 2, 1, 3 and 3, 1, 2 in Cases 1 and 2, respectively. The snapshots of the initial positions and final fencing formations for both cases are provided in Fig. 5.7c through f, illustrating the effectiveness of the long-term target fencing framework in Eq. (5.15).

Moreover, we focus on Fig. 5.7a to quantitatively analyze the state evolution. As illustrated in Fig. 5.8, the fencing errors $e = [e_x, e_y]^\top \in \mathbb{R}^3$ from Eq. (5.48) converge to near zero by $t = 20$ s. This demonstrates that the virtual target is effectively fenced by the three AMRs, fulfilling Objective (1) in Definition 5.3. The distances between robots and between each robot and the target, shown in Fig. 5.8, satisfy $\|x_{2,1}(t)\| > 1.5$, $\|x_{1,3}(t)\| > 1.5$, $\|x_{3,2}(t)\| > 1.5, \forall t > 0$, which prevents overlapping as required by Objective (3) in Definition 5.3. Figure 5.8 reveals that the relative distances between adjacent robots, following the ordering sequence $s[1] - s[3] :=$

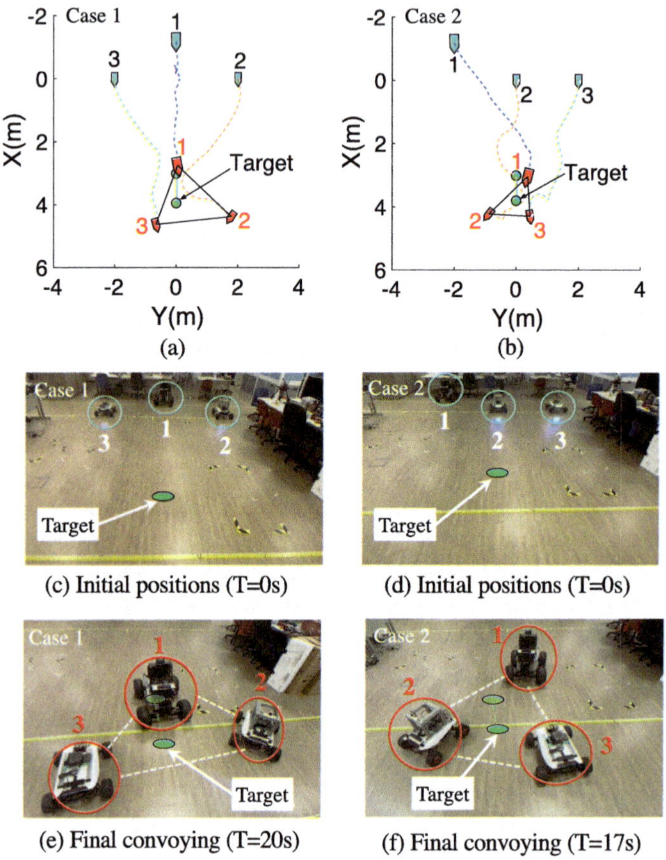

Fig. 5.7 Two experimental cases of *ordering-flexible* target fencing with different initial positions. Subfigures **a**, **b**: Trajectories of three AMRs starting from different initial positions form a fencing formation with distinct spatial orderings in Cases 1-2. Subfigures **c**, **d**: Snapshots of initial positions of the three AMRs in Cases 1–2. Subfigures **e**, **f**: Snapshots of the final stable fencing formation formed by the three AMRs in Cases 1–2. (Here, the blue and red vehicles in subfigures **a**, **b** denote the initial and the final positions of the three AMRs, respectively. The green circle denotes the moving target. Moreover, the larger AMR Hunter 1.0 is labeled 1, whereas the other two Scout Mini are specified with labels 2, 3)

$\{2, 1, 3\}$, fall within the range $1.5 < \lim_{t\to\infty} \|x_{s[i],s[i+1]}(t)\| < 4$. This ensures compliance with Objective (2) in Definition 5.3. Finally, Fig. 5.9 indicates that the desired velocities $u_i = [u_{i,x}, u_{i,y}]^\top \in \mathbb{R}^3$, $i = 1, 2, 3$, for each robot i are consistently within the bound $\zeta = 0.2$, i.e., $\|u_i(t)\|_\infty \leq 0.2$ for all $t > 0$, which confirms the feasibility of the long-term target fencing described in Eq. (5.15).

Remark 5.14 The velocity-tracking module used in the 2D experiments is provided by AGILE·X Company. This module is proprietary, so public users only have

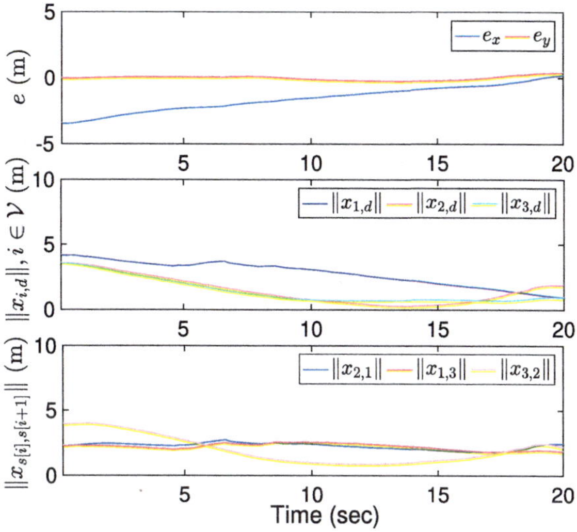

Fig. 5.8 Temporal evolution of the fencing errors $e = [e_x, e_y]^\top \in \mathbb{R}^2$, the robot-target distances $\|x_{i,d}\|$, $i = 1, 2, 3$, and the relative distances between adjacent robots $\|x_{2,1}\|$, $\|x_{1,3}\|$, $\|x_{3,2}\|$ with an ordering sequence in a 2D plane in Fig. 5.7a for example

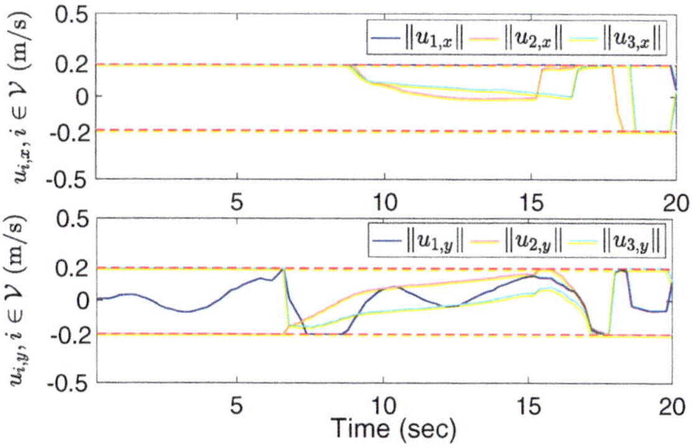

Fig. 5.9 Temporal evolution of the AMR desired velocities $u_i := [u_{i,x}, u_{i,y}]^\top \in \mathbb{R}^2$, $i = 1, 2, 3$, in Fig. 5.7a for example. (The dashed red lines are the upper and lower limits of the input)

access to transmit desired command velocities to the AMRs, without insight into the detailed tracking methods. Before using this module for fencing experiments, we conducted several tests to verify its performance. The AMRs were able to track both constant and varying velocities with only minimal disturbances. These tiny disturbances are further mitigated by the outer-loop target fencing described in Eq. (5.15), as the velocity-tracking module operates within the inner loop of the closed-loop system. Thus, the ordering-flexible target fencing experiments can bc effectively conducted with the encapsulated velocity-tracking module. It is important to note that the proposed framework in Eq. (5.15) does not directly handle frequent transient changes in the commanded velocities $u_i, i \in \mathcal{V}$. If the velocities $u_i, i \in \mathcal{V}$, experience transient changes, their derivatives may become infinite at some time $t = T_f > 0$ (i.e., $\lim_{t \to T_f} \dot{u}_i = \infty$). This would contradict the boundedness of the states $\ddot{\Phi}_i$ and compromise the convergence of convex-hull fencing as established in Lemma 5.3. Consequently, handling such transient behaviors will be the focus of future research.

5.4.2 3D Simulations

In this subsection, we perform 3D simulations to assess the validity of Theorem 5.1 in the context of varying environmental factors within a higher-dimensional Euclidean space. We analyze a multi-robot system comprising six robots, controlled by the dynamics outlined in Eqs. (5.5) and (5.15), with an input limit set to $\zeta_i = 6$. For consistency, we use the same collision radius r, sensing radius R, and the parameters η_1 and η_2 as specified in Sect. 5.4.1. In the following experiments, we examine *ordering-flexible* multi-robot fencing with two types of moving targets: one exhibiting linear motion and the other circular motion.

For the line-motion target, we assume the target moves at a constant velocity $v_d := [1, 0, 0]^\top$ m/s, tracked by a similar velocity estimator as described in Sect. 5.4.1. Figure 5.10a, b show the trajectories of six robots starting from various initial positions (blue triangles) fulfilling Assumption 5.5, leading to a hexahedron-pattern target fencing (red triangles) with two different orderings: $\{1, 4, 6, 5, 2, 3\}$ and $\{3, 5, 1, 6, 4, 2\}$ in 3D Euclidean space. For the circular-motion target, we set the target's velocity as $v_d := [2\omega \cos(\theta + \pi/2), 2\omega \sin(\theta + \pi/2, 0]^\top$ m/s, where θ represents the angle between the target x_d and the centroid of the desired circle $[6, 10, 0]^\top$, and $\omega = 0.1\,\mathrm{s}^{-1}$ is the angular velocity. Figure 5.10c, d depict the trajectories of six robots starting from the same initial positions as in Sect. 5.4.1, leading to the desired hexahedron-pattern fencing (red triangles) with orderings $\{1, 3, 6, 4, 2, 5\}$ and $\{3, 5, 1, 6, 4, 2\}$. These simulations validate the flexible-ordering property in Objective 2 of Definition 5.2.

Additionally, using Fig. 5.10b as an example, we analyze the state evolution of the *ordering-flexible* multi-robot fencing in 3D space. As shown in Fig. 5.11, the fencing errors $e = [e_x, e_y, e_z]^\top \in \mathbb{R}^3$ across all three axes converge to zero, indicating that the target is eventually fenced by the six robots, fulfilling Objective (1) in Definition 5.2. The robot-target distances $\|x_{i,d}\|, i \in \mathcal{V}$, and the inter-robot distances $\|x_{i,j}\|, i \neq j \in \mathcal{V}$, in Fig. 5.11 satisfy $\|x_{i,j}(t)\| \geq 1.5$, $\|x_{i,d}(t)\| > 0$ for all $i \neq j \in \mathcal{V}$ and $t > 0$, which thus prevents overlapping behavior as required by Objective (3) of Definition 5.2. Furthermore, Fig. 5.12 shows that the relative distances between adjacent robots in the ordering sequence $\{3, 5, 1, 6, 4, 2\}$ satisfies $1.5 < \lim_{t\to\infty} \|x_{s[i],s[i+1]}(t)\| < 4$, $s[1] - s[6] := \{3, 5, 1, 6, 4, 2\}$, ensuring Objective (2) of Definition 5.2. Finally, Fig. 5.13 demonstrates that the control inputs $u_i = [u_{i,x}, u_{i,y}, u_{i,z}]^\top \in \mathbb{R}^3, i \in \mathcal{V}$, for robot i are consistently bounded by $\zeta = 6$, i.e., $\|u_i(t)\|_\infty \leq 6$ for all $t > 0$, ensuring the feasibility of the long-term target fencing (5.15).

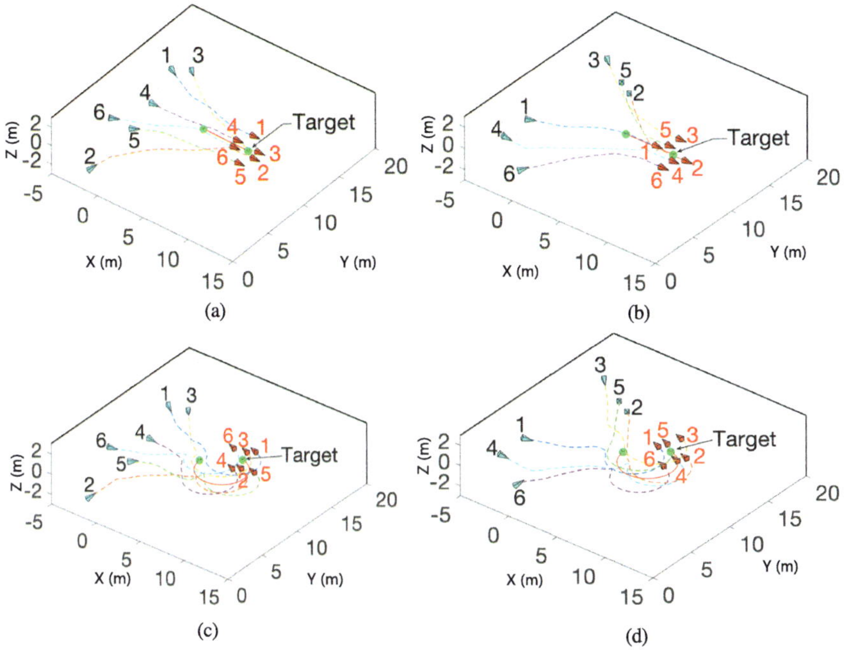

Fig. 5.10 Four cases of a six-robot system from two different initial positions to the *ordering-flexible* hexahedron-pattern fencing for a target of different dynamics using the proposed constraint-based optimization framework (5.15) in 3D. Subfigures **a**, **b**: A line-motion *ordering-flexible* target fencing (i.e., a constant velocity $v_d := [1, 0, 0]^\top$ m/s). Subfigures **c**, **d**: A circular-motion *ordering-flexible* target fencing (i.e., a variational velocity $v_d := [2\omega\cos(\theta + \pi/2), 2\omega\sin(\theta + \pi/2, 0]^\top$ m/s) with θ being the relative angle between the target x_d and the centroid of desired circle, ω the corresponding angular velocity. (Here, the blue and red triangles represent the initial and final positions of the robots, respectively. The green ball and red line denote the target and its trajectory, respectively)

To demonstrate the robustness of the *ordering-flexible* target fencing approach, Fig. 5.14 shows scenarios where two robots fail unexpectedly at $t = 0.5s$ during the six-robot target fencing task. Specifically, as depicted in Fig. 5.14a, the remaining four robots ($i = 1, 2, 3, 5$) successfully maintain tetrahedron-pattern fencing with the ordering $\{5, 3, 1, 2\}$ even after robots $i = 4$ and $i = 6$ fail. Similarly, in Fig. 5.14b, the robots $i = 1, 3, 4, 6$ achieve the desired tetrahedron-pattern fencing with the ordering $\{3, 1, 4, 6\}$ when robots $i = 2$ and $i = 5$ break down. These results confirm the robustness of the LTTE algorithm in 3D, demonstrating a capability that surpasses previous methods [4–14].

Furthermore, to evaluate the obstacle-avoidance capability, we introduce two static spherical obstacles into the six-robot system as described in Remark 5.13. The positions and radii of these obstacles are specified as $x_1^o = [2, 8, 0]^\top$ m with $r_1^o = 1.2$ m and $x_2^o = [2, 13, 0]^\top$ m with $r_2^o = 1$ m, ensuring that no robots are initially inside them. Figure 5.15 illustrates the trajectories of the six robots starting from the same initial positions as in Fig. 5.10b (blue triangles), transitioning through

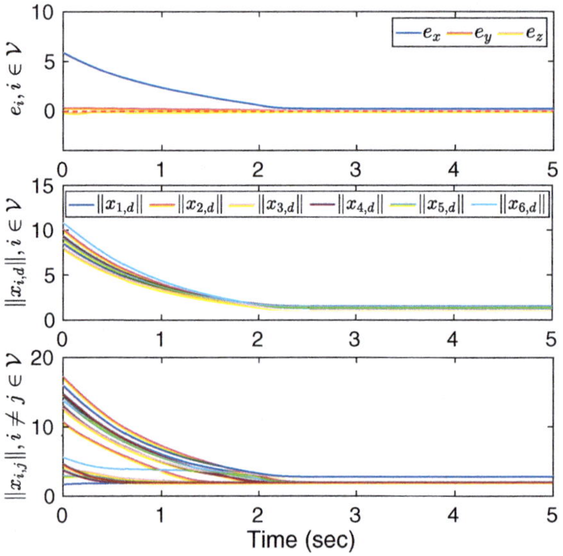

Fig. 5.11 Temporal evolution of the fencing errors $e = [e_x, e_y, e_z]^\top \in \mathbb{R}^3$, the robot-target distances $\|x_{i,d}\|$, $i \in \mathcal{V}$, and the inter-robot distances $\|x_{i,j}\|$, $i \neq j \in \mathcal{V}$ in Fig. 5.10b for example

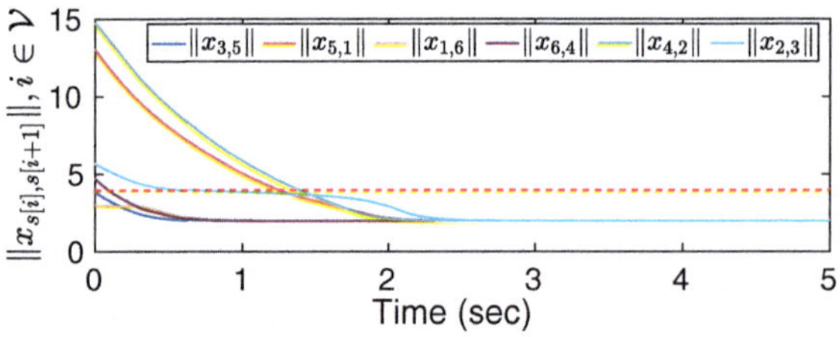

Fig. 5.12 Temporal evolution of the relative distances between adjacent robots in the ordering sequence $\|x_{3,5}\|$, $\|x_{5,1}\|$, $\|x_{1,6}\|$, $\|x_{6,4}\|$, $\|x_{4,2}\|$ in Fig. 5.10b for example

obstacle avoidance (yellow triangles), and finally forming the hexahedron-pattern fencing (red triangles). As shown in Fig. 5.15a, the dashed trajectories demonstrate that all six robots successfully avoid the obstacles (dark blue spheres), confirming that the obstacle-avoidance requirement is met. Figure 5.15b provides a clearer view of the final hexahedron-pattern fencing from another perspective. Additionally, Fig. 5.16 shows that the fencing errors $e = [e_x, e_y, e_z]^\top \in \mathbb{R}^3$ converge to zero, and the distances between the robots and obstacles satisfy $x_{i,1}^o(t) > 1.2$, $x_{i,2}^o(t) > 1$ for all $i \in \mathcal{V}$ and $t > 0$, thereby confirming the obstacle-avoidance capability mentioned in Remark 5.13.

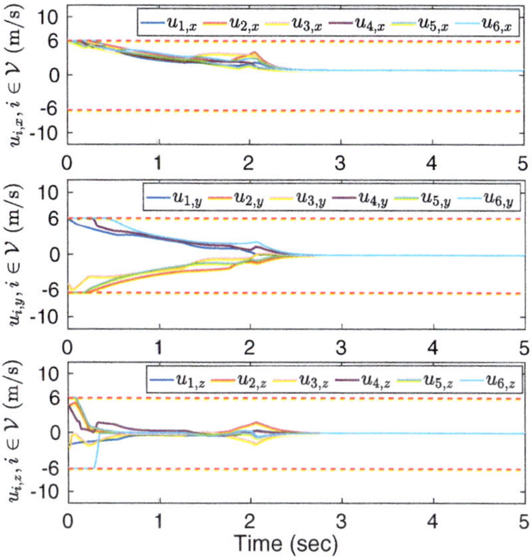

Fig. 5.13 Temporal evolution of the robot inputs $u_i := [u_{i,x}, u_{i,y}, u_{i,z}]^\top \in \mathbb{R}^3$, $i \in \mathcal{V}$, in Fig. 5.10b for example. (The dashed red lines are the upper and lower limits of the input)

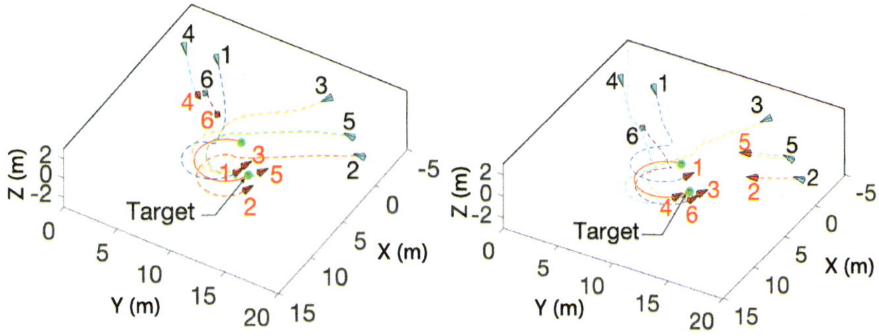

(a) (b)

Fig. 5.14 Special situations of two robots suddenly breakdown at $t = 0.5$ s in the six-robot *ordering-flexible* target fencing task. Subfigure **a**: The rest four robots $i = 1, 2, 3, 5$ form the successful tetrahedron-pattern fencing when robots $i = 4, 6$ break down. Subfigure **b**: The rest four robots $i = 1, 3, 4, 6$ form the successful tetrahedron-pattern fencing when robots $i = 2, 5$ break down. (All the symbols have the same meaning as in Fig. 5.10)

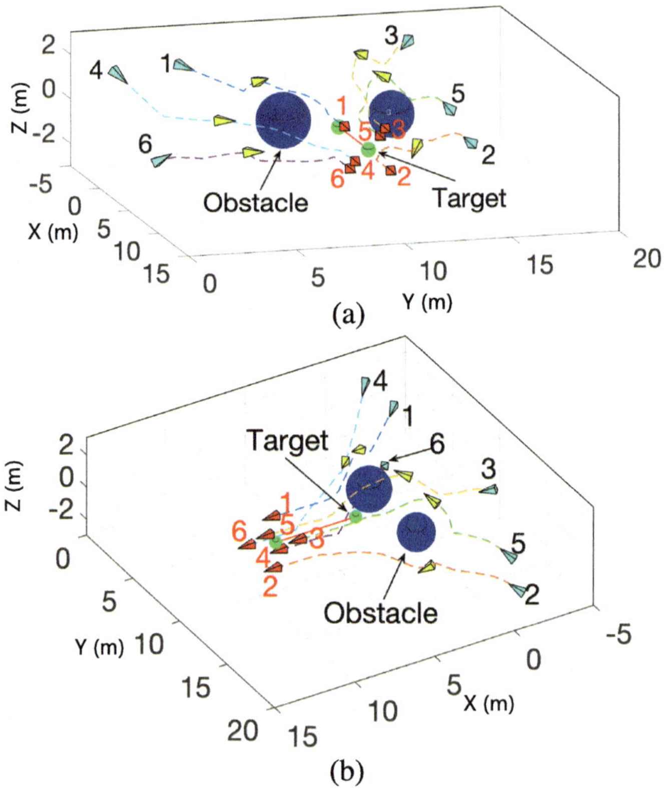

Fig. 5.15 **a** Trajectories of six robots and the target from the same initial positions in Fig. 5.10b with two static ball obstacles. **b** Another view to better illustrate the desired hexahedron-pattern fencing. (Here, the yellow triangles represent the obstacle-encountering positions of robots. All the other symbols have the same meaning as in Fig. 5.10)

5.5 Conclusion

In this chapter, we have proposed an LTTE algorithm that enables *ordering-flexible* multi-robot target fencing by integrating fencing subtasks as constraints within an online constraint-based optimization framework. These constraints facilitate the execution of long-term target-fencing tasks in dynamic environments. The algorithm's global convergence is rigorously analyzed, accounting for time-varying collision avoidance among neighbors and external disturbances that decay exponentially. The effectiveness, adaptability across multiple dimensions, and robustness of the LTTE approach are validated through 2D *ordering-flexible* fencing experiments with three AMRs and 3D simulations.

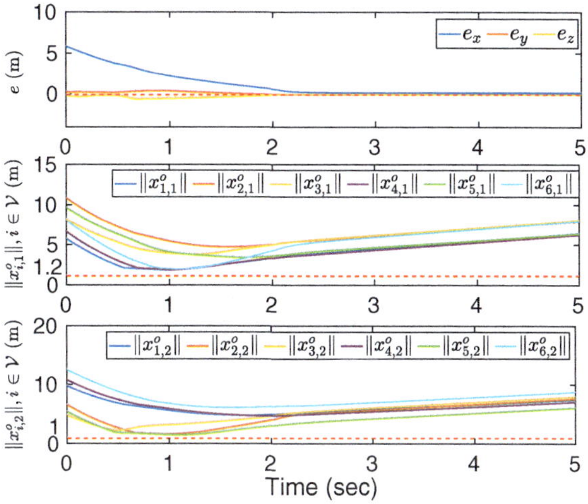

Fig. 5.16 Temporal evolution of the fencing errors $e = [e_x, e_y, e_z]^\top \in \mathbb{R}^3$, and the robot-obstacles distances $x^o_{i,1}, x^o_{i,2}, i \in \mathcal{V}$, in Fig. 5.14b for example

References

1. J. Han, J. Zhang, C. He, C. Lv, X. Hou, Y. Ji, Distributed finite-time safety consensus control of vehicle platoon with sensor and actuator failures. IEEE Trans. Veh. Technol. **72**(1), 162–175 (2022)
2. B.-B. Hu, H.-T. Zhang, B. Liu, H. Meng, G. Chen, Distributed surrounding control of multiple unmanned surface vessels with varying interconnection topologies. IEEE Trans. Control Syst. Technol. **30**(1), 400–407 (2021)
3. M.J. Islam, J. Mo, J. Sattar, Robot-to-robot relative pose estimation using humans as markers. Auton. Robots **45**(4), 579–593 (2021)
4. F. Chen, W. Ren, Y. Cao, Surrounding control in cooperative agent networks. Syst. Control Lett. **59**(11), 704–712 (2010)
5. B. Hu, H. Zhang, Bearing-only motional target-surrounding control for multiple unmanned surface vessels. IEEE Trans. Ind. Electron. **69**(4), 3988–3997 (2021)
6. Y.J. Shi, R. Li, K.L. Teo, Cooperative enclosing control for multiple moving targets by a group of agents. Int. J. Control **88**(1), 80–89 (2015)
7. B.-B. Hu, H.-T. Zhang, J. Wang, Multiple-target surrounding and collision avoidance with second-order nonlinear multiagent systems. IEEE Trans. Ind. Electron. **68**(8), 7454–7463 (2020)
8. K. Sakurama, H.-S. Ahn, Multi-agent coordination over local indexes via clique-based distributed assignment. Automatica **112**, 108670 (2020)
9. B. Liu, Z. Chen, H.-T. Zhang, X. Wang, T. Geng, H. Su, J. Zhao, Collective dynamics and control for multiple unmanned surface vessels. IEEE Trans. Control Syst. Technol. **28**(6), 2540–2547 (2019)
10. C. Wang, G. Xie, Limit-cycle-based decoupled design of circle formation control with collision avoidance for anonymous agents in a plane. IEEE Trans. Control Syst. Technol. **62**(12), 6560–6567 (2017)
11. B.-B. Hu, Z. Chen, H.-T. Zhang, Distributed moving target fencing in a regular polygon formation. IEEE Trans. Control Netw. Syst. **9**(1), 210–218 (2022)
12. L. Kou, Z. Chen, J. Xiang, Cooperative fencing control of multiple vehicles for a moving target with an unknown velocity. IEEE Trans. Autom. Control **67**(2), 1008–1015 (2021)

13. L. Kou, Y. Huang, Z. Chen, S. He, J. Xiang, Cooperative fencing control of multiple second-order vehicles for a moving target with and without velocity measurements. Int. J. Robust Nonlinear Control **31**(10), 4602–4615 (2021)
14. B.-B. Hu, H.-T. Zhang, Y. Shi, Cooperative Label-free moving target fencing for second-order multi-agent systems with rigid formation. Automatica **148**, 110788 (2023)
15. G. Notomista, M. Egerstedt, Constraint-driven coordinated control of multi-robot systems, in *Proceedings of the American Control Conference (ACC)* (2019), pp. 1990–1996
16. G. Notomista, S. Mayya, Y. Emam, C. Kroninger, A. Bohannon, S. Hutchinson, M. Egerstedt, A resilient and energy-aware task allocation framework for heterogeneous multirobot systems. IEEE Trans. Robot. **38**(1), 159–179 (2021)
17. W. Yao, H.G. de Marina, B. Lin, M. Cao, Singularity-free guiding vector field for robot navigation. IEEE Trans. Robot. **37**(4), 1206–1221 (2021)
18. M. Egerstedt, J.N. Pauli, G. Notomista, S. Hutchinson, Robot ecology: Constraint-based control design for long duration autonomy. Annu. Rev. Control **46**, 1–7 (2018)
19. D. Mellinger, V. Kumar, Minimum snap trajectory generation and control for quadrotors, in *Proceedings of IEEE International Conference on Robotics and Automation (ICRA)* (2011), pp. 2520–2525
20. S. Ruder, An overview of gradient descent optimization algorithms (2016). arXiv:1609.04747 (2016)
21. L. Wang, A.D. Ames, M. Egerstedt, Safety barrier certificates for collisions-free multirobot systems. IEEE Trans. Robot. **33**(3), 661–674 (2017)
22. A.D. Ames, X. Xu, J.W. Grizzle, P. Tabuada, Control barrier function based quadratic programs for safety critical systems. IEEE Trans. Autom. Control **62**(8), 3861–3876 (2016)
23. X. Zhou, X. Wen, Z. Wang, Y. Gao, H. Li, Q. Wang, T. Yang, H. Lu, Y. Cao, C. Xu et al., Swarm of micro flying robots in the wild. Sci. Robot. **7**(66), eabm5954 (2022)
24. B.-B. Hu, H.-T. Zhang, W. Yao, J. Ding, M. Cao, Spontaneous-ordering platoon control for multirobot path navigation using guiding vector fields. IEEE Trans. Robot. **39**(4), 2654–2668 (2023)
25. J. Wu, Z. Huang, Z. Hu, C. Lv, Toward human-in-the-loop AI: enhancing deep reinforcement learning via real-time human guidance for autonomous driving. Engineering **21**, 75–91 (2023)
26. B.-B. Hu, H. Zhang, B. Liu, J. Ding, Y. Xu, C. Luo, H. Cao, Coordinated navigation control of cross-domain unmanned systems via guiding vector fields. IEEE Trans. Control Syst. Technol. **32**(2), 550–563 (2024)
27. Z. Chen, Synchronization of frequency modulated multi-agent systems. IEEE Trans. Autom. Control **68**(6), 3425–3439 (2023)
28. H. Wei, K. Zhang, H. Zhang, Y. Shi, Resilient and constrained consensus against adversarial attacks: a distributed MPC framework. Automatica **160**, 111417 (2024)
29. J. Huang, *Nonlinear Output Regulation: Theory and Applications* (SIAM, Philadelphia, USA, 2004)
30. A. Isidori, *Nonlinear Control Systems II* (Springer, London, UK, 2013)
31. Y. Hong, J. Hu, L. Gao, Tracking control for multi-agent consensus with an active leader and variable topology. Automatica **42**(7), 1177–1182 (2006)
32. S. Boyd, L. Vandenberghe, *Convex Optimization* (Cambridge University Press, UK, 2004)
33. N.J. Higham, *Accuracy and Stability of Numerical Algorithms* (SIAM, UK, 2004)
34. B.N. Parlett, *The Symmetric Eigenvalue Problem.* (SIAM, 1998)
35. K.H. Khalil, *Nonlinear Systems* (Prentice Hall, Upper Saddle River, 2002)
36. P. Glotfelter, I. Buckley, M. Egerstedt, Hybrid nonsmooth barrier functions with applications to provably safe and composable collision avoidance for robotic systems. IEEE Robot. Autom. Lett. **4**(2), 1303–1310 (2019)
37. M.F. Reis, A.P. Aguiar, P. Tabuada, Control barrier function-based quadratic programs introduce undesirable asymptotically stable equilibria. IEEE Control Syst. Lett. **5**(2), 731–736 (2020)
38. T. Qin, J. Pan, S. Cao, S. Shen, A general optimization-based framework for local odometry estimation with multiple sensors (2019). arXiv:1901.03638

Part II
Multiple-Target Scenario

Chapter 6
Coordinated Multiple-Target Fencing Control and Collision Avoidance

6.1 Introduction

The previous three Chaps. 3–5 focused on the theme of "complex fencing," and successively proposed cooperative fencing controllers that are based on input-to-state stability, output regulation theory, and long-term task execution, respectively. Although they covered various functions such as target fencing, pattern regulation, and neighboring collision avoidance, the aforementioned methods are only applicable to a single target, which has not yet been studied with multiple varying-velocity targets. In practical scenarios involving the collaborative fencing of multiple unmanned surface vehicles (USVs), there may be some situations requiring the fencing for multiple targets of varying velocities. However, fencing multiple targets with varying velocities raises a series of new issues, such as how to reasonably define the fencing of multiple varying-velocity targets and the topological connectivity between targets and robots. Therefore, this chapter continues to focus on the theme of "complex fencing" and aims to study the problem of cooperative fencing control for multiple varying-velocity targets. In what follows, we will introduce the related background.

Over the decades, significant advancements have been made in the cooperative control of MRSs [1–5], spanning fields such as biology, physics, systems science, computer engineering, and swarm intelligence. Investigating the coordination mechanisms of MRSs sheds light on the dynamics of biological collective behavior and provides insights for analyzing and designing networked scientific and engineering systems. These mechanisms play a crucial role in applications like wireless sensor networks, smart grids, wind farm networks, and unmanned systems, enhancing cooperative efficiency, adaptability to system uncertainties, and robustness against external disturbances.

Significant efforts have thus far been focused on developing and analyzing consensus and flocking protocols for MRSs [6–10]. In recent years, growing attention has been directed toward containment control, which aims to guide robots into the convex hull defined by target positions [11–13]. Moreover, fencing control, which

B.-B. Hu and H.-T. Zhang, *Multi-Robot Systems*, Cognitive Intelligence and Robotics,
https://doi.org/10.1007/978-981-97-9214-6_6

drives a team of robots to move around target(s) within a convex hull, has emerged as the reverse problem of containment control. Both containment and fencing strategies apply to tasks such as collaborative escort, and patrolling with USVs, as well as formation control for UAVs, UGVs, and satellite or spacecraft clusters.

Due to the inherent challenges in stabilizing the collective behavior of robots around multiple targets, most existing research has concentrated on single-target scenarios [14–20]. Early studies primarily addressed single static targets. For example, Lan et al. [14] introduced a hybrid control strategy for cyclically enclosing a static target. Liu and Chen [15] proposed a hierarchical control structure with output regulation to achieve target fencing. Later, Chen [16] tackled a fencing problem with ensured collision avoidance. As target dynamics became more complex in real-world marine applications, attention shifted toward fencing control for moving targets. For instance, Lin and Dong [17] developed a robust real-time vision system for target tracking, while Guo et al. [18] presented a distributed control method for surrounding a moving target based on relative neighbor positions. Zhang et al. [19] utilized a game-theory framework to track a moving target while maintaining a specific distance. Yu and Liu [20] devised a dynamic distributed control law to form circular formations of unicycles with switching interconnection topologies, with the target accessible to only one unicycle. Subsequently, Li et al. [21] introduced a cooperative fencing scheme with collision avoidance guarantees for networked MRSs with nonlinear Lagrangian dynamics.

The primary difficulty in addressing the multiple-target fencing control problem is the need for robots to estimate the center of multiple moving targets without direct access to their information. Among the limited research on this topic, Chen and Ren [22] focused on enclosing a group of stationary targets and proposed a fencing control law within an estimation-and-control framework with a fixed topology. Following this work, Shoja et al. [23] introduced an adaptive approach to tackle the fencing control of non-identical robots with unknown system parameters. Most earlier studies have concentrated on first-order MRSs due to the complexities involved in controller design. However, many real-world systems, such as unmanned vehicles and sensor networks, have at least second-order dynamics. To address this, Shi et al. [24] explored the collective fencing problem for linear second-order MRSs using LaSalle's Invariance Principle [25], considering both stationary and moving targets. Despite these contributions, current research on multiple-target fencing control is largely confined to simulations of linear MRSs, with little progress in real-world experiments involving actual robots, which limits the practical application of these potentially effective protocols.

As a key example of nonlinear second-order systems, USVs are increasingly vital for marine activities such as resource exploration, rescue operations, naval attacks, and water area surveillance. While some studies [15, 26] have addressed collective fencing control for single target vessels, the development of fencing protocols for multiple target vessels is still ongoing. This is due to the complexities of designing distributed state estimator-based controllers and the difficulties of implementing these protocols in real-world scenarios. Furthermore, the challenge of collision avoidance adds to the difficulty. With the growing demands from the modern marine industry and the need for advanced resource exploration, it has become essential and advantageous to develop a systematic distributed protocol for controlling multiple moving targets with collision avoidance. In response to this need, this chapter presents a collective approach for second-order nonlinear MRSs to fence multiple moving targets with guaranteed collision avoidance. The main contributions of this chapter are two-fold:

1. Design a quasi-equal-distance fencing strategy for a second-order nonlinear MRS to fence multiple moving targets while ensuring collision avoidance, utilizing a state estimator to track the center of the targets.
2. Implement this distributed fencing control strategy in our established indoor multi-USV system to validate its effectiveness.

The rest of this chapter is organized as follows: Sect. 6.2 provides preliminaries and outlines the main problem. Sections 6.3–6.5 present a fencing strategy for encircling multiple targets and derive the conditions for asymptotic stability. Section 6.6 discusses experimental results to validate the effectiveness of the proposed method. Finally, conclusions are drawn in Sect. 6.7.

6.2 Problem Formulation of Fencing Multiple Varying-Velocity Targets

In this section, we consider an MRS consisting of n follower robots: $I_v = \varsigma_i \mid i = 1, 2, \ldots, n$, where ς_i denotes the follower robot i with the following second-order dynamics:

$$\begin{cases} \dot{x}_i(t) = v_i(t), \\ \dot{v}_i(t) = f_i(v_i, t) + u_i(t), \end{cases} \tag{6.1}$$

where $i \in I_v$, $x_i(t), v_i(t) \in \mathbb{R}^2$ represent the position and velocity of follower robot i, respectively, $u_i(t) \in \mathbb{R}^2$ is the control input, and $f_i(v_i, t) \in \mathbb{R}^2$ is a nonlinear function assumed to be of the form:

$$f_i(v_i, t) = \phi_i^{\mathsf{T}}(v_i, t)\theta_i. \tag{6.2}$$

where $\phi_i(\cdot) : \mathbb{R}^2 \times \mathbb{R} \to \mathbb{R}^{p\times 2}$ is a nonlinear vector function of the input v_i, t and $\theta_i \in \mathbb{R}^p$, $i \in I_v$ constant coefficients. Moreover, there exists m targets $I_t = \{\tau_j | j = n+1, n+2, \ldots, n+m\}$ with τ_j set as target j in the MRS, each of which moving with a variable velocity described as,

$$\dot{r}_j(t) = v_j(t), \tag{6.3}$$

where $r_j(t), v_j(t) \in \mathbb{R}^2$ represent the position and velocity of target j, respectively. Assume that $v_j(t)$ in Eq. (6.3) can be expressed as

$$v_j(t) = \phi_L^{\mathsf{T}}(t)\eta_j, \tag{6.4}$$

where $\phi_L(\cdot) : \mathbb{R} \to \mathbb{R}^{p\times 2}$ is a nonlinear vector function of t with constant coefficients $\eta_j \in \mathbb{R}^p$, $j \in I_t$.

For the multi-robot-multi-target MRS, let $\mathcal{G}(\mathcal{V}, \mathcal{E})$ be the communication graph, where $\mathcal{V} = 1, 2, \ldots, n+m = I_v \cup I_t$ represents the vertex set, and $\mathcal{E} = (\mathcal{V}_i, \mathcal{V}_k) \subseteq \mathcal{V} \times \mathcal{V}$ denotes the edge set. Here, $(\mathcal{V}_i, \mathcal{V}_k)$ indicates that vertex $\mathcal{V}_i$ is a neighbor of vertex $\mathcal{V}_k$. The connections within the follower set I_v are undirected if, for any two vertices $i, k \in I_v$, both $(\mathcal{V}_i, \mathcal{V}_k) \in \mathcal{E}$ and $(\mathcal{V}_k, \mathcal{V}_i) \in \mathcal{E}$ hold. Let $\mathcal{N}_i \subseteq I_v$ denote the neighbor(s) of follower robot i. The matrix $B = [b_{ij}] \in \mathbb{R}^{n\times m}$, where $i \in I_v$ and $j \in I_t$, represents the connection between the follower robot set I_v and the target set I_t. In this matrix, $b_{ij} = 1$ if target j is a neighbor of follower robot i, and $b_{ij} = 0$ otherwise. The undirected connection within the follower set is considered connected if there is at least one path between any two vertices $\mathcal{V}_i, \mathcal{V}_k \in I_v$. The following assumptions are made for further derivations.

Assumption 6.1 The connection $\mathcal{G}(\mathcal{V}, \mathcal{E})$ among the follower robot set I_v is assumed to be undirected and connected. Furthermore, the connection between the target set I_t and the follower robot set I_v satisfies the following conditions: $\sum_{i=1}^{n} b_{ij} = 1, \sum_{j=n+1}^{n+m} b_{ij} \leq 1, i \in I_v, j \in I_t$.

Under Assumption 6.1, each target in I_t is linked to exactly one unique follower robot in I_v, and each follower robot in I_v can be connected to no more than one target in I_t. Connections between multiple targets and a single follower, or multiple followers and a single target, are not permitted. It is important to note that the practical significance of Assumption 6.1 is its role in estimating the center of targets when only partial information about the targets is available.

Lemma 6.1 ([23]) *Under Assumption 6.1, one has*

$$b_{ij} \sum_{k=n+1}^{n+m} b_{ik} = b_{ij},\ j, k \in I_t.$$

Additionally, it is assumed that the distance between each target and the center of I_t is bounded, as specified in the following assumption.

Assumption 6.2 There exists $M \in \mathbb{R}^+$ such that

$$M = \sup\{\|r_j(t) - \bar{r}(t)\|\},\ j \in I_t,$$

with $\bar{r}(t) := \sum_{j=n+1}^{n+m} r_j(t)/m$.

With Assumptions 6.1 and 6.2, we can give the following definition.

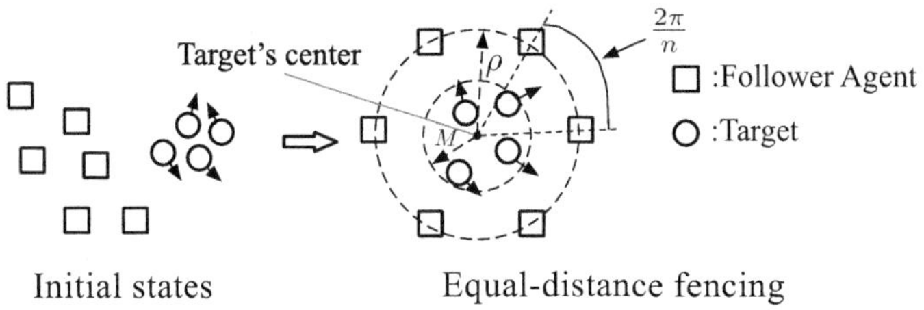

Fig. 6.1 A equal-distance fencing scenario for a four-target six-robot MRS

Definition 6.1 (*Equal-distance fencing* [15]) A set of m targets is asymptotically surrounded by n follower robots if each follower robot $i \in I_v$ satisfies

$$\lim_{t\to\infty} \|x_i(t) - \bar{r}(t) - \rho_i\| = 0, \tag{6.5}$$

where $\rho_i = \left[\rho\cos\frac{2\pi i}{n}, \rho\sin\frac{2\pi i}{n}\right]^{\top} \in \mathbb{R}^2$ is the relative fencing vector for follower robot $i \in I_v$, with $\rho > M$ representing a constant radius.

Definition 6.1 indicates that the follower robots are arranged at equal distances around the center of the targets, as shown in Fig. 6.1. Additionally, if $\rho > M$ is suitably selected, the set of robots I_v will effectively encircle the set of targets I_t, thus addressing the equal-distance fencing control problem.

Definition 6.2 (*Quasi-equal-distance fencing*) A set of m targets is asymptotically fenced by n follower robots if each follower robot $i \in I_v$ satisfies

$$\lim_{t\to\infty} \|x_i(t) - \bar{r}(t) - \rho_i\| \le \sigma, \tag{6.6}$$

with the threshold $\sigma > 0$, quasi-equal-distance fencing is achieved.

Obviously, Definition 6.2 is more practical and less stringent than the equal-distance fencing described in Definition 6.1.

Definition 6.3 (*Collision avoidance*) For any two follower robots $i, k \in I_v$, the collision is avoided if

$$\|x_{i,k}(t)\| > r, t \ge 0,$$

where $x_{i,k} := x_i - x_k$ and $r > 0$ is a safe distance.

It is important to note that the targets have varying velocities and are assumed to be free of collisions. With this in mind, we can now present the main technical issues addressed in this chapter.

Problem 6.1 Design a control signal

$$u_i = f(x_i, x_k, r_j), i \in I_v, k \in N_i, j \in I_t$$

for robot i in a multi-target multi-robot MRS $I_v \cup I_t$, governed by Eqs. (6.1) and (6.3), to achieve a quasi-equal-distance fencing configuration.

Problem 6.2 Design a control signal

$$u_i = f(x_i, x_k, r_j), i \in I_v, k \in \mathcal{N}_i, j \in I_t$$

to address Problem 6.1 while ensuring guaranteed collision avoidance (as defined in Definition 6.3) among the follower robot set I_v.

6.3 Estimation of the Center of Targets and Analysis

Let $\widehat{x}_i(t)$ and $\widehat{v}_i(t)$, for $i \in I_v$, denote the estimated center position and velocity of the target set for follower robot i, respectively, as given below

$$\begin{cases} \dot{\widehat{x}}_i(t) = \widehat{v}_i(t), \\ \dot{\omega}_i(t) = \beta_p \sum\limits_{j \in \mathcal{N}_i} \mathrm{sig}\big(\widehat{x}_j(t) - \widehat{x}_i(t)\big), \\ \widehat{x}_i(t) = \omega_i(t) + \widetilde{r}_i(t). \end{cases} \tag{6.7}$$

Here, $\omega_i(t)$ represents the consensus part with an initial state of $\omega_i(0) = 0$, $\beta_p \in \mathbb{R}^+$ is the estimator gain, and $\widetilde{r}_i(t) = \frac{n}{m}\sum_{j=n+1}^{n+m} b_{ij} r_j(t),\ i \in I_v,\ j \in I_t$ is derived from the target group. It follows from Eq. (6.7) that

$$\frac{1}{n}\sum_{i=1}^{n} \widehat{x}_i(t) = \frac{1}{n}\sum_{i=1}^{n} \widetilde{r}_i(t) = \frac{1}{m}\sum_{i=1}^{n}\sum_{j=n+1}^{n+m} b_{ij} r_j(t) = \bar{r}(t) \tag{6.8}$$

by the virtue of Assumption 6.1.

Lemma 6.2 ([27]) *For the estimator governed by Eq. (6.7), it follows that*

$$\frac{d\|e_p\|^2}{dt} \le \sum_{i=1}^{n}\sum_{j \in \mathcal{N}_i} \{\|\widehat{x}_j - \widehat{x}_i\|((n-1)\alpha_1 - \beta_p\|\widehat{x}_j - \widehat{x}_i\|)\},$$

where

$$\|e_p\|^2 = \sum_{i=1}^{n} (\widehat{x}_i - \bar{r})^2 \tag{6.9}$$

is the estimated position error and $\alpha_1 = \sup\{\|\dot{\widetilde{r}}_i(t)\| \;|i \in I_v, t \ge 0\}$.

To estimate the center state $\bar{r}_i(t)$ and $\dot{\bar{r}}_i(t)$ of the target set I_t, an additional assumption is required.

Assumption 6.3 There exist upper bounds of $\|\dot{\tilde{r}}_i(t)\|$, $\|\ddot{\tilde{r}}_i(t)\|$, $\|\dddot{\tilde{r}}_i(t)\|$, $i \in I_v$ defined in Eq. (6.7).

Note that $|\dot{\tilde{r}}_i(t)|$, $|\ddot{\tilde{r}}_i(t)|$, and $|\dddot{\tilde{r}}_i(t)|$ for $i \in I_v$ are accessible to the followers with the aid of Assumption 6.1. In practice, Assumption 6.3 implies that the motional targets do not undergo sharp braking or turning, which corresponds to the bounded inter-distance among the targets.

Lemma 6.3 *For the positive real numbers σ_p and $\sigma_a \in \mathbb{R}^+$, the estimates of position $\widehat{x}_i$ and acceleration $\dot{\widehat{v}}_i$ converge to the center of I_t under Assumption 6.3, adhering to specified bounds, such that*

$$\sum_{i=1}^{n}(\widehat{x}_i - \bar{r})^2 \leq \sigma_p^2, \sum_{i=1}^{n}(\dot{\widehat{v}}_i - \ddot{\bar{r}})^2 \leq \sigma_a^2$$

provided that there exists an estimated gain $\beta_p > 0$ in Eq. (6.7) satisfying

$$\beta_p > \bar{\beta} \tag{6.10}$$

with $\bar{\beta} = \max\left\{2(n-1)\alpha_1, 2(n-1)\alpha_2, \frac{(n-1)^2\alpha_1}{2\sigma_p}, \frac{(n-1)^2\alpha_2}{2\sigma_a}, \sqrt[3]{\frac{(n^2-2)(n-1)^3\alpha_1^2}{4\sigma_p}}, \sqrt[3]{\frac{(n^2-2)(n-1)^3\alpha_2^2}{4\sigma_a}}\right\}$, *where* $\alpha_1 := \sup\{\|\dot{\tilde{r}}_i(t)\| \,\big|\, i \in I_v, t \geq 0\}$ *and* $\alpha_2 := \sup\{\|\dddot{\tilde{r}}_i(t)\| \,\big|\, i \in I_v, t \geq 0\}$.

Proof Let a Lyapunov function be $V_1(t) = (\|e_p\|^2 - \sigma_p^2)^2$, whose derivative is

$$\dot{V}_1(t) = 2(\|e_p\|^2 - \sigma_p^2)\frac{d\|e_p\|^2}{dt}.$$

According to Lemma 6.2, one has

$$\frac{d\|e_p\|^2}{dt} \leq \sum_{i=1}^{n}\sum_{j\in\mathcal{N}_i}\{\|\widehat{x}_j - \widehat{x}_i\|((n-1)\alpha_1 - \beta_p\|\widehat{x}_j - \widehat{x}_i\|)\} \tag{6.11}$$

with $\alpha_1 = \sup\{\|\dot{\tilde{r}}_i(t)\| \,\big|\, i \in I_v, t \geq 0\}$. Moreover, in accordance to Eq. (6.8), one has

$$\|e_p\|^2 = \sum_{i=1}^{n}(\widehat{x}_i - \bar{r})^2 = \sum_{i=1}^{n}\left(\widehat{x}_i - \frac{1}{n}\sum_{j=1}^{n}\widehat{x}_j\right)^2.$$

It can be derived that

$$\|e_p\| = \sqrt{\sum_{i=1}^{n}(\widehat{x}_i - \frac{1}{n}\sum_{j=1}^{n}\widehat{x}_j)^2} \leq \sum_{i=1}^{n}\|\widehat{x}_i - \frac{1}{n}\sum_{j=1}^{n}\widehat{x}_j\| \leq (n-1)\max\{\|\widehat{x}_i - \widehat{x}_j\|, i, j \in I_v\},$$

which implies $\max\{\|\widehat{x}_i - \widehat{x}_j\|, i, j \in I_v\} > \frac{\sigma_p}{n-1}$ if $\|e_p\|^2 > \sigma_p^2$. Furthermore, if

$$\frac{(n-1)\alpha_1}{2\beta_p} < \frac{\sigma_p}{n-1}\left(i.e.,\ \beta_p > \frac{(n-1)^2\alpha_1}{2\sigma_p}\right),$$

one has the following inequality

$$\begin{aligned}\frac{d\|e_p\|^2}{dt} <&(n^2-2)\frac{(n-1)^2\alpha_1^2}{4\beta_p^2} + 2\frac{\sigma_p}{n-1}\left((n-1)\alpha_1 - \beta_p\frac{\sigma_p}{n-1}\right),\\ <&-\frac{2\sigma_p}{n-1}\beta_p + \frac{(n^2-2)(n-1)^2\alpha_1^2}{4\beta_p^2} + 2\sigma_p\alpha_1.\end{aligned}$$

with the virtue of Eq. (6.11). Note that

$$f_p(\beta_p) = -\frac{2\sigma_p}{n-1}\beta_p + \frac{(n^2-2)(n-1)^2\alpha_1^2}{4\beta_p^2} + 2\sigma_p\alpha_1$$

is monotonically decreasing with respect to β_p, which implies that

$$\begin{aligned}\frac{d\|e_p\|^2}{dt} <& -\frac{\sigma_p}{n-1}\beta_p + \frac{(n^2-2)(n-1)^2\alpha_1^2}{4\beta_p^2} - \frac{\sigma_p}{n-1}\beta_p + 2\sigma_p\alpha_1\\ <& -\frac{2\sigma_p}{n-1}\beta_p + \frac{(n^2-2)(n-1)^2\alpha_1^2}{4\beta_p^2} + 2\sigma_p\alpha_1\\ <& \ 0,\end{aligned}$$

if $\beta_p > \max\left\{2(n-1)\alpha_1, \sqrt[3]{\frac{(n^2-2)(n-1)^3\alpha_1^2}{4\sigma_p}}, \frac{(n-1)^2\alpha_1}{2\sigma_p}\right\}$ is picked. Accordingly, the derivative $\dot{V}_1(t) < 0$ if $\|e_p\|^2 > \sigma_p^2$ and $\frac{d\|e_p\|^2}{dt} < 0$, which implies that $V_1(t)$ decreases if $\|e_p\|^2 \geq \sigma_p^2$. Thereby, $\|e_p\|$ eventually converges into a set $\{e_p \mid \|e_p\|^2 \leq \sigma_p^2\}$.

We will next demonstrate the boundedness of $\sum_{i=1}^n(\dot{\widehat{v}}_i - \ddot{\bar{r}})^2$. To begin, we consider the dynamics of the estimated acceleration for the center of the targets to be

$$\dot{\widehat{v}}_i(t) = \ddot{\omega}_i(t) + \ddot{\bar{r}}_i(t), i \in I_v,$$

with $\ddot{\omega}_i(t) = \beta_p \sum_{j\in N_i}(\widehat{v}_j - \widehat{v}_i)$. Note that

$$\|e_a\|^2 = \sum_{i=1}^n\left(\dot{\widehat{v}}_i - \frac{1}{n}\sum_{j=1}^n\dot{\widehat{v}}_j\right)^2,$$

one thus has $\|e_a\|^2 = \sum_{i=1}^{n}(\hat{\dot{v}}_i - \ddot{r})^2$ according to Eq. (6.8). Then, let a Lyapunov function be $V_2(t) = (\|e_a\|^2 - \sigma_a^2)^2$ with $\sigma_a > 0$, one has

$$\dot{V}_2(t) = 2(\|e_a\|^2 - \sigma_a^2)\frac{d\|e_a\|^2}{dt},$$

and

$$\frac{d\|e_a\|^2}{dt} < \sum_{i=1}^{n}\sum_{j\in\mathcal{N}_i}\left\{|\hat{\dot{v}}_j - \hat{\dot{v}}_i|((n-1)\alpha_2 - \beta_p|\hat{\dot{v}}_j - \hat{\dot{v}}_i|)\right\}$$

with $\alpha_2 = \sup\{|\dddot{r}_i(t)|\ \big|i \in I_v, t \geq 0\}$. If $\|e_a\|^2 > \sigma_a^2$, the following inequality holds $\frac{\sigma_a}{n-1} < \max\{|\hat{\dot{v}}_i - \hat{\dot{v}}_j||i, j \in I_v\}$. Furthermore, if $\frac{(n-1)\alpha_2}{2\beta_p} < \frac{\sigma_a}{n-1}$ (i.e., $\beta_p > \frac{(n-1)^2\alpha_2}{2\sigma_a}$), one has

$$\frac{d\|e_a\|^2}{dt} < -\frac{2\sigma_a}{n-1}\beta_p + \frac{(n^2-2)(n-1)^2\alpha_2^2}{4\beta_p^2} + 2\sigma_a\alpha_2 < 0,$$

if $\beta_p > \max\{2(n-1)\alpha_2, \sqrt[3]{\frac{(n^2-2)(n-1)^3\alpha_2^2}{4\sigma_a}}, \frac{(n-1)^2\alpha_2}{2\sigma_a}\}$ is picked. It thus follows that $\dot{V}_2(t) < 0$ when $\|e_a\|^2 > \sigma_a^2$ and $\frac{d\|e_a\|^2}{dt} < 0$. Therefore, $\|e_p\|$, $\|e_a\|$ will eventually converge into the set $\{e_p, e_a\big|\ \|e_a\| \leq \sigma_a, \|e_p\| \leq \sigma_p\}$ if $\beta_p > \bar{\beta}$ is picked with $\bar{\beta} = \max\left\{2(n-1)\alpha_1, 2(n-1)\alpha_2, \frac{(n-1)^2\alpha_1}{2\sigma_p}, \frac{(n-1)^2\alpha_2}{2\sigma_a}, \sqrt[3]{\frac{(n^2-2)(n-1)^3\alpha_1^2}{4\sigma_p}},\right.$ $\left.\sqrt[3]{\frac{(n^2-2)(n-1)^3\alpha_2^2}{4\sigma_a}}\right\}$. The proof is thus completed. □

Remark 6.1 The convergence speed of the estimation errors e_p and e_a is influenced by the estimator gain β_p in Eq. (6.7). A sufficiently large value of β_p can help mitigate the effects of the upper boundary of the target's dynamic variations. Furthermore, $\bar{\beta}$ indicates the practical upper limit of the dynamic variations of the targets.

6.4 Equal-Distance Fencing and Convergence Analysis

We are now set to introduce a distributed fencing controller to tackle Problem 6.1. Before presenting the specifics of the controller, we first establish the following assumption regarding θ_i and η_j for $i \in I_v$ and $j \in I_t$ as used in Eqs. (6.2) and (6.4).

Assumption 6.4 There exist positive real numbers θ_M and η_M such that for any follower robot $i \in I_v$ and any target $j \in I_t$, the following holds: $|\theta_i| \leq \theta_M$ and $|\eta_j| \leq \eta_M$.

The Assumption 6.4 is reasonable in practical scenarios, the nonlinear function $f_i(v_i, t)$ and the target velocities are bounded, ensuring the existence of θ_M and

η_M. These upper bounds on the uncertain parameters are critical for understanding and controlling the fencing errors.

Prior to introducing the control protocol, let us define the equal-distance fencing error $e_i(t)$ for the ith follower robot. This definition uses the estimator provided in Eq. (6.7) and is expressed as below

$$e_i(t) := x_i(t) - \widehat{x}_i(t) - \rho_i, \tag{6.12}$$

where $i \in I_v$ and ρ_i are given in Definition 6.1. Accordingly,

$$\begin{cases} \dot{e}_i = \delta_i, \\ \dot{\delta}_i = u_i + \phi_i^{\mathrm{T}}\theta_i - \dot{\widehat{v}}_i, \end{cases} \tag{6.13}$$

where $\delta_i := v_i - \widehat{v}_i$ is the velocity error between the follower robot i and the estimated center $\widehat{v}_i$ of I_t. From Eq. (6.13), it immediately follows that

$$\begin{cases} \dot{e}_i = -k_{i1}e_i + z_i, \\ \dot{z}_i = u_i + \phi_i^{\mathrm{T}}\theta_i - \dot{\widehat{v}}_i + k_{i1}z_i - k_{i1}^2 e_i \end{cases} \tag{6.14}$$

with $z_i := \delta_i + k_{i1}e_i$. An adaptive controller is thus proposed as below

$$\begin{aligned} u_i &= k_{i1}^2 e_i - k_{i2}z_i - \phi_i^{\mathrm{T}}\widehat{\theta}_i + \sum_{j=n+1}^{n+m} b_{ij}\dot{\phi}_L^{\mathrm{T}}(t)\widehat{\eta}_j^i, \\ \dot{\widehat{\theta}}_i &= \phi_i z_i - k_{i3}\widehat{\theta}_i, \\ \dot{\widehat{\eta}}_j^i &= -\dot{\phi}_L z_i - k_{i4}\sum_{k=n+1}^{n+m} b_{ik}\hat{\eta}_k^i, \end{aligned} \tag{6.15}$$

where $k_{i1}, k_{i2}, k_{i3}, k_{i4} \in \mathbb{R}^+,\ i \in I_v$, are the control gains of robot i to be designed afterwards.

Theorem 6.1 *Given Assumptions 6.1 through 6.4, the closed-loop MRS governed by the dynamics in Eqs. (6.1), (6.3), and (6.14), along with the controller specified in Eq. (6.15), is able to achieve quasi-equal-distance fencing.*

Proof Let $\widetilde{\theta}_i, \widetilde{\eta}_j^i$ be the parameter errors to be

$$\widetilde{\theta}_i := \widehat{\theta}_i - \theta_i,\ \widetilde{\eta}_j^i := \sum_{k=n+1}^{n+m} b_{ik}\widehat{\eta}_k^i - \eta_j \tag{6.16}$$

with $i \in I_v$, $j, k \in I_t$. The first part of (6.16) implies that $\dot{\widetilde{\theta}}_i = \dot{\widehat{\theta}}_i$. By Assumption 6.1, it is obtained that $b_{ij}b_{ik} = b_{ij}$, if $j = k$, and $b_{ij}b_{ik} = 0$, otherwise, with $k, j \in I_t$. Then, the derivative of $\widetilde{\eta}_j^i$ in (6.16) becomes

$$\dot{\widetilde{\eta}}_j^i = \sum_{k=n+1}^{n+m} b_{ik}\dot{\widehat{\eta}}_k^i, i \in I_v, \ \ j,k \in I_t.$$

Consider Lemma 6.1, one has

$$b_{ij}\dot{\widetilde{\eta}}_j^i = b_{ij} \sum_{k=n+1}^{n+m} b_{ik}\dot{\widehat{\eta}}_k^i = b_{ij}\dot{\widehat{\eta}}_j^i. \tag{6.17}$$

Substituting (6.15) and (6.17) into (6.16) yields

$$\begin{aligned}\dot{\widetilde{\theta}}_i &= \phi_i z_i - k_{i3}\widehat{\theta}_i, \\ \dot{\widetilde{\eta}}_j^i &= -\dot{\phi}_L z_i - k_{i4} \sum_{k=n+1}^{n+m} b_{ik}\widehat{\eta}_k^i.\end{aligned} \tag{6.18}$$

Let a Lyapunov function be

$$V(t) = V_1(t) + V_2(t) + V_3(t), \tag{6.19}$$

where

$$V_1(t) := \frac{1}{2}\sum_{i=1}^{n}\{e_i^{\mathsf{T}} e_i + z_i^{\mathsf{T}} z_i\}, \ V_2(t) := \frac{1}{2}\sum_{i=1}^{n}\widetilde{\theta}_i^{\mathsf{T}}\widetilde{\theta}_i, \ V_3(t) := \frac{1}{2m}\sum_{i=1}^{n}\sum_{j=n+1}^{n+m}(\widetilde{\eta}_j^i)^{\mathsf{T}}\widetilde{\eta}_j^i.$$

It thus follows from Eq. (6.16) that

$$\begin{aligned}2\widehat{\theta}_i^{\mathsf{T}}\widetilde{\theta}_i &\geq \widetilde{\theta}_i^{\mathsf{T}}\widetilde{\theta}_i - \theta_i^{\mathsf{T}}\theta_i, \\ 2\sum_{k=n+1}^{n+m} b_{ik}(\widehat{\eta}_k^i)^{\mathsf{T}}\widetilde{\eta}_j^i &\geq (\widetilde{\eta}_j^i)^{\mathsf{T}}\widetilde{\eta}_j^i - {\eta_j}^{\mathsf{T}}\eta_j.\end{aligned}$$

Accordingly, one has the derivative of $V_i(t), i = 1, 2, 3$ fulfill

$$\begin{aligned}\dot{V}_1(t) \leq& \sum_{i=1}^{n}\Big\{-\Big(k_{i1} - \frac{1}{2}\Big)e_i^{\mathsf{T}} e_i + \frac{1}{2}z_i^{\mathsf{T}} z_i - (k_{i2} - k_{i1})z_i^{\mathsf{T}} z_i + z_i^{\mathsf{T}}(-\phi_i^{\mathsf{T}}\widetilde{\theta}_i + \sum_{j=n+1}^{n+m} b_{ij}\dot{\phi}_L^{\mathsf{T}}\widehat{\eta}_j^i - \dot{\widehat{v}}_i\}, \\ \dot{V}_2(t) \leq& \sum_{i=1}^{n}\Big\{z_i^{\mathsf{T}}\phi_i^{\mathsf{T}}\widetilde{\theta}_i - \frac{k_{i3}}{2}\widetilde{\theta}_i^{\mathsf{T}}\widetilde{\theta}_i + \frac{k_{i3}}{2}\theta_i^{\mathsf{T}}\theta_i\Big\}, \\ \dot{V}_3(t) \leq& -\sum_{i=1}^{n} z_i^{\mathsf{T}}\dot{\phi}_L^{\mathsf{T}}\sum_{k=n+1}^{n+m} b_{ik}\widehat{\eta}_k^i + \frac{1}{m}\sum_{i=1}^{n}\sum_{j=n+1}^{n+m} z_i^{\mathsf{T}}\dot{\phi}_L^{\mathsf{T}}\eta_j - \frac{k_{i4}}{2m}\sum_{i=1}^{n}\sum_{j=n+1}^{n+m}(\widetilde{\eta}_j^i)^{\mathsf{T}}\widetilde{\eta}_j^i \\ &+ \frac{k_{i4}}{2m}\sum_{i=1}^{n}\sum_{j=n+1}^{n+m}\eta_j^{\mathsf{T}}\eta_j,\end{aligned}$$

which thus implies that

$$\dot{V}(t) \leq \sum_{i=1}^{n}\Big\{-\Big(k_{i1}-\frac{1}{2}\Big)e_i^{\mathsf{T}}e_i-(k_{i2}-k_{i1}-1)z_i^{\mathsf{T}}z_i-\frac{k_{i3}}{2}\widetilde{\theta}_i^{\mathsf{T}}\widetilde{\theta}_i-\frac{k_{i4}}{2m}\sum_{j=n+1}^{n+m}(\widetilde{\eta}_j^i)^{\mathsf{T}}\widetilde{\eta}_j^i\Big\}$$
$$+\sum_{i=1}^{n}\Big\{\frac{k_{i3}}{2}\theta_i^{\mathsf{T}}\theta_i+\frac{k_{i4}}{2m}\sum_{j=n+1}^{n+m}\eta_j^{\mathsf{T}}\eta_j+\frac{1}{2}\kappa_i^{\mathsf{T}}\kappa_i\Big\} \tag{6.20}$$

with $\kappa_i = \frac{1}{m}\sum_{j=n+1}^{n+m}\dot{\phi}_L^{\mathsf{T}}\eta_j - \dot{\widehat{v}}_i$.

Furthermore, it follows from Eq. (6.7) that $\sum_{i=1}^{n}\beta_p\sum_{j\in N_i}\mathrm{sig}\big(\widehat{x}_j(t)-\widehat{x}_i(t)\big)=0$. Considering Eq. (6.8), one has $\frac{1}{n}\sum_{i=1}^{n}\widehat{v}_i=\frac{1}{m}\sum_{j=n+1}^{n+m}\phi_L^{\mathsf{T}}\eta_j$, which implies that $\frac{1}{n}\sum_{i=1}^{n}\dot{\widehat{v}}_i=\frac{1}{m}\sum_{j=n+1}^{n+m}\dot{\phi}_L^{\mathsf{T}}\eta_j$. Therefore, it follows from Lemma 6.3 that

$$\sum_{i=1}^{n}\Big(\frac{1}{m}\sum_{j=n+1}^{n+m}\dot{\phi}_L^{\mathsf{T}}\eta_j-\dot{\widehat{v}}_i\Big)^2=\sum_{i=1}^{n}\Big(\frac{1}{n}\sum_{j=1}^{n}\dot{\widehat{v}}_j-\dot{\widehat{v}}_i\Big)^2\leq\sigma_a^2,$$

which implies that $\sum_{i=1}^{n}\frac{1}{2}\kappa_i^{\mathsf{T}}\kappa_i\leq\frac{\sigma_a^2}{2}$. Then, Eq. (6.20) directly leads to that

$$\dot{V}(t)\leq-\sum_{i=1}^{n}\Big\{\Big(k_{i1}-\frac{1}{2}\Big)e_i^{\mathsf{T}}e_i+(k_{i2}-k_{i1}-1)z_i^{\mathsf{T}}z_i+\frac{k_{i3}}{2}\widetilde{\theta}_i^{\mathsf{T}}\widetilde{\theta}_i+\frac{k_{i4}}{2m}\sum_{j=n+1}^{n+m}(\widetilde{\eta}_j^i)^{\mathsf{T}}\widetilde{\eta}_j^i\Big\}+\epsilon_*$$

with

$$\epsilon_*=\sum_{i=1}^{n}\left\{\frac{k_{i3}}{2}\theta_i^{\mathsf{T}}\theta_i+\frac{k_{i4}}{2m}\sum_{j=n+1}^{n+m}\eta_j^{\mathsf{T}}\eta_j\right\}+\frac{\sigma_a^2}{2}.$$

It is clear that

$$\epsilon_*\leq\epsilon_M=\sum_{i=1}^{n}\left\{\frac{k_{i3}}{2}\theta_M^2+\frac{k_{i4}}{2m}\sum_{j=n+1}^{n+m}\eta_M^2\right\}+\frac{\sigma_a^2}{2}$$

according to Assumption 6.4. Let

$$m_{i1}:=k_{i1}-\frac{1}{2}>0,\ m_{i2}:=k_{i2}-k_{i1}-1>0,$$
$$m_{i3}:=\frac{k_{i3}}{2}>0,\ m_{i4}:=\frac{k_{i4}}{2m}>0, \tag{6.21}$$

and define γ as

$$\gamma=[e^{\mathsf{T}},z^{\mathsf{T}},\widetilde{\theta}^{\mathsf{T}},\widetilde{\eta}^{\mathsf{T}}]^{\mathsf{T}} \tag{6.22}$$

with $e := [e_1^{\mathsf{T}}, e_2^{\mathsf{T}}, \ldots, e_n^{\mathsf{T}}]^{\mathsf{T}}$, $z := [z_1^{\mathsf{T}}, z_2^{\mathsf{T}}, \ldots, z_n^{\mathsf{T}}]^{\mathsf{T}}$, $\widetilde{\theta} := [\widetilde{\theta}_1^{\mathsf{T}}, \widetilde{\theta}_2^{\mathsf{T}}, \ldots, \widetilde{\theta}_n^{\mathsf{T}}]^{\mathsf{T}}$, $\widetilde{\eta} := [(\widetilde{\eta}_1^1)^{\mathsf{T}}, (\widetilde{\eta}_1^2)^{\mathsf{T}}, \ldots, (\widetilde{\eta}_1^n)^{\mathsf{T}}, \ldots, (\widetilde{\eta}_m^1)^{\mathsf{T}}, (\widetilde{\eta}_m^2)^{\mathsf{T}}, \ldots, (\widetilde{\eta}_m^n)^{\mathsf{T}}]^{\mathsf{T}}$, it is derived that

$$\dot{V}(t) \leq -\gamma^{\mathsf{T}} Q \gamma + \epsilon_M$$

with $Q := \operatorname{diag}\{m_{11}, m_{21}, \ldots, m_{n1}, m_{12}, m_{22}, \ldots, m_{n2}, m_{13}, m_{32}, \ldots, m_{n3}, m_{14} \otimes 1_{\mathbf{m}}^{\mathsf{T}}, m_{24} \otimes 1_{\mathbf{m}}^{\mathsf{T}}, \ldots, m_{n4} \otimes 1_{\mathbf{m}}^{\mathsf{T}}\} \succ 0$. Accordingly, $\gamma^{\mathsf{T}} Q \gamma \geq \lambda_{\min}(Q)\|\gamma\|^2$ with $\lambda_{\min}(Q) > 0$. Then, it is obtained that

$$\dot{V}(t) \leq -\lambda_{\min}(Q)\|\gamma\|^2 + \epsilon_M.$$

From Eq. (6.19), one has that $V(t) = \gamma^{\mathsf{T}} P \gamma$, where

$$P = \operatorname{diag}\Bigg\{\underbrace{\frac{1}{2}, \frac{1}{2}, \ldots, \frac{1}{2}}_{3n}, \underbrace{\frac{1}{2m}, \frac{1}{2m}, \ldots, \frac{1}{2m}}_{n\times m}\Bigg\} \succ 0$$

with $\lambda_{\max}(P) = 1/2$ and $\lambda_{\min}(P) = 1/2m$. It immediately leads to

$$\dot{V}(t) \leq -\lambda_{\min}(Q)\|\gamma\|^2 + \epsilon_M \leq -2\lambda_{\min}(Q)V(t) + \epsilon_M.$$

Direct calculation gives

$$V(t) \leq V(0)e^{-\int_0^{\mathsf{T}} \xi ds} + e^{-\int_0^{\mathsf{T}} \xi ds}\int_0^t \epsilon_M e^{\xi s} ds \leq V(0)e^{-\xi t} + \frac{\epsilon_M}{\xi}(1 - e^{-\xi t})$$

with $\xi = 2\lambda_{\min}(Q)$. In accordance to definition of $V(t)$ in Eq. (6.19), one has

$$\lim_{t\to\infty} \|\gamma\| \leq \lim_{t\to\infty} \sqrt{\frac{V(t)}{\lambda_{\min}(P)}} \leq \sqrt{\frac{m\epsilon_M}{\lambda_{\min}(Q)}},$$

which implies that e_i, z_i, θ_i, η_j^i, $i \in I_v$, $j \in I_t$, are all bounded by γ defined in Eq. (6.22). Taking into account the boundedness of the estimated position error $|e_p|$ as established by Lemma 6.3, the proof is thereby concluded. □

Remark 6.2 The equal-distance fencing error $e_i(t)$ for $i \in I_v$, as defined in Eq. (6.12), can be minimized by adjusting the upper boundary $\sqrt{m\epsilon_M/\lambda_{\min}(Q)}$, which is controlled by the gains $k_{i1}, k_{i2}, k_{i3}, k_{i4}$. This boundary also satisfies the inequality in Eq. (6.21). Similarly, the estimation errors for $\widetilde{\theta}_i$ (for $i \in I_v$) and $\widetilde{\eta}_j$ (for $j \in I_t$) can be managed, effectively reducing the boundary ϵ_M.

6.5 Equal-Distance Fencing with Collision Avoidance and Convergence Analysis

Next, we will tackle Problem 6.2. Define a potential function $U_{i,k}$ between two follower robots i and k as follows

$$U_{i,k} := \begin{cases} 0 & , \ \|x_{i,k}\| > R, \\ \left(\frac{\|x_{i,k}\|^2 - R^2}{\|x_{i,k}\|^2 - r^2}\right)^2, & r < \|x_{i,k}\| \le R, \end{cases} \tag{6.23}$$

where R represents the radius of the detection region and r is the safe radius defined in Definition 6.3. Clearly, $U_{i,k}$ approaches ∞ as $|x_i - x_k| \to r$. Taking the partial derivative of $U_{i,k}$ with respect to x_i, we obtain

$$\frac{\partial U_{i,k}}{\partial x_i} = \begin{cases} 0 & , \ \|x_{i,k}\| > R, \\ x_{i,k}^{\mathrm{T}} \frac{4(R^2 - r^2)(\|x_{i,k}\|^2 - R^2)}{(\|x_{i,k}\|^2 - r^2)^3} & , \ r < \|x_{i,k}\| \le R. \end{cases} \tag{6.24}$$

Note that

$$\lim_{\|x_{i,k}\| \to R^-} U_{i,k} = \lim_{\|x_{i,k}\| \to R^+} U_{i,k} = 0, \quad \lim_{\|x_{i,k}\| \to R^-} \frac{\partial U_{i,k}}{\partial x_i} = \lim_{\|x_{i,k}\| \to R^+} \frac{\partial U_{i,k}}{\partial x_i} = 0$$

with R^-, R^+ representing the left and right limit of R. Apparently, both $U_{i,k}$ and $\frac{\partial U_{i,k}}{\partial x_i}$ are continuous for $\|x_{i,k}\| > r$. It follows from Eq. (6.24) that

$$\frac{\partial U_{i,k}}{\partial x_i} = -\frac{\partial U_{i,k}}{\partial x_k} = \frac{\partial U_{k,i}}{\partial x_i} = -\frac{\partial U_{k,i}}{\partial x_k}.$$

Analogously, an adaptive fencing control law with collision avoidance is designed to be

$$\begin{aligned} u_i =& -k_{i1}e_i - k_{i2}\delta_i - \phi_i^{\mathrm{T}}\hat{\theta}_i + \dot{\hat{v}}_i - k_{i3}\sum_{k=1}^{n}\frac{\partial U_{i,k}}{\partial x_k}, \\ \dot{\hat{\theta}}_i =& \phi_i\delta_i, \end{aligned} \tag{6.25}$$

where $k_{i1}, k_{i2}, k_{i3} \in \mathbb{R}^+$ are the control gains for follower robot i, and e_i and δ_i are defined in Eq. (6.13), respectively. The last term on the first line of Eq. (6.25) is responsible for collision avoidance.

To continue with the analysis, we provide the following assumption.

Assumption 6.5 Initially, for each pair of follower robots i and k, it holds that $|x_{i,k}(0)| > r$, where $i, k \in I_v$.

Assumption 6.5 is essential for ensuring collision avoidance. It mandates that the follower robots begin from a configuration where no collisions occur. With this assumption in place, we can now present the main technical result.

Theorem 6.2 *Under Assumptions 6.1–6.3, 6.5, the closed-looped MRS with dynamics (6.1), (6.3), (6.13), (6.24), and (6.25) is able to achieve quasi-equal-distance fencing with guaranteed collision avoidance.*

Proof (i) (Collision avoidance) We will prove collision avoidance by contradiction. Let a Lyapunov function be

$$V(t) = V_1(t) + V_2(t) + V_3(t), \tag{6.26}$$

where

$$V_1(t) = \frac{1}{2}\sum_{i=1}^{n}\{k_{i1}e_i^{\mathrm{T}}e_i + \delta_i^{\mathrm{T}}\delta_i\},\ V_2(t) = \frac{1}{2}\sum_{i=1}^{n}\widetilde{\theta}_i^{\mathrm{T}}\widetilde{\theta}_i,\ V_3(t) = \frac{1}{2}\sum_{i=1}^{n}\sum_{k=1}^{n}k_{i3}U_{i,k}\,.$$

It follows from Eq. (6.24) that

$$\sum_{i=1}^{n}\sum_{k=1}^{n}\hat{v}_i^{\mathrm{T}}\frac{\partial U_{i,k}}{\partial x_i} = \hat{v}_i^{\mathrm{T}}\sum_{i=1}^{n}\sum_{k=1}^{n}\frac{\partial U_{i,k}}{\partial x_i} = 0, \tag{6.27}$$

and

$$\begin{aligned}
\dot{V}_1(t) &= \sum_{i=1}^{n}\left\{-k_{i1}\delta_i^{\mathrm{T}}e_i - k_{i2}\delta_i^{\mathrm{T}}\delta_i + \delta_i^{\mathrm{T}}\left(-\phi_i^{\mathrm{T}}\widetilde{\theta}_i - k_{i5}\sum_{k=1}^{n}\frac{\partial U_{i,k}}{\partial x_k}\right)\right\},\\
\dot{V}_2(t) &= \sum_{i=1}^{n}(\phi_i\delta_i)^{\mathrm{T}}\widetilde{\theta}_i,\\
\dot{V}_3(t) &= \frac{1}{2}\sum_{i=1}^{n}\sum_{k=1}^{n}k_{i3}\left(\dot{x}_i^{\mathrm{T}}\frac{\partial U_{i,k}}{\partial x_i} + \dot{x}_k^{\mathrm{T}}\frac{\partial U_{i,k}}{\partial x_k}\right), = \dot{x}_i^{\mathrm{T}}\sum_{i=1}^{n}\sum_{k=1}^{n}k_{i3}\frac{\partial U_{i,k}}{\partial x_i} = \delta_i^{\mathrm{T}}\sum_{i=1}^{n}\sum_{k=1}^{n}k_{i3}\frac{\partial U_{i,k}}{\partial x_i}.
\end{aligned}$$

Accordingly, it together with Eq. (6.26) that

$$\dot{V}(t) = -\sum_{i=1}^{n}k_{i2}\delta_i^{\mathrm{T}}\delta_i \le 0,$$

which implies that $V(t)$ is non-increasing and bounded for all $t \in [0, \infty)$. Additionally, we have

$$\frac{1}{2}\sum_{i=1}^{n}\sum_{k=1}^{n}k_{i3}U_{i,k} \le V(t),$$

which indicates that $U_{i,k}$, for any pair of follower robots i and k in I_v, is bounded. This is inconsistent with $\lim_{|x_{i,k}(t)|\to r^+} U_{i,k} = \infty$, where r^+ denotes the limit approaching r. Therefore, $\|x_{i,k}\|$ must stay greater than r, preventing any collisions among the follower robots in I_v.

(ii) (Quasi-equal-distance fencing)

In part (i), we have demonstrated that $V(t)$ is uniformly bounded. Consequently, this implies that e_i, δ_i, $\tilde{\theta} i$, and Ui, k for $i \in I_v$ are all uniformly bounded. Additionally, the fencing position error $e_i(t)$ remains bounded. According to Lemma 6.3, the estimated position error $\|e_p\|$ for the targets' center is also bounded. This resolves Problem 6.2, which completes the proof. □

Remark 6.3 The equal-distance fencing problem can also be addressed with the availability of the acceleration $\hat{\dot{v}}_i(t)$ of the estimated center (6.7). The proof for this is analogous to that of Theorem 6.2 and is therefore omitted here for brevity.

6.6 Algorithm Verification

In this section, we present experiments on our indoor multi-USV system, focusing on quasi-equal-distance fencing with multiple moving targets while ensuring collision avoidance.

6.6.1 A Multi-USV Experiments Platform

Figure 6.2a shows our indoor multi-USV system, which features a USV motion capture system (MCS), a control station (CS), six 300 mm-long 3D-printed HUSTER-0.3 vessels, and a 3000 mm×4000 mm pool. The MCS is equipped with eight infrared cameras and optical motion capture software, allowing for the creation of a spatial Cartesian coordinate system in the pool and precise measurement of vessel positions with an accuracy of ±3 mm. Each vessel includes two DC motors, speed encoders, an onboard wireless device, infrared emitters, indicators, and an LI-PO battery, all managed by a microcontroller unit (MCU).

As depicted in Fig. 6.2b, the system employs an infrared emitter mounted on each vessel along with the USV motion capture system to monitor the relative positions and velocities of the vessels. This data is transmitted to the control station, which then communicates with each HUSTER-0.3 vessel over a wireless network operating at 100 Hz. Each vessel uses this information to generate control signals aimed at achieving quasi-equal-distance fencing. Furthermore, the control station tracks the status trajectories and controller parameters to analyze performance. It is important to note that the control strategy is decentralized, with each vessel only interacting with its neighboring vessels.

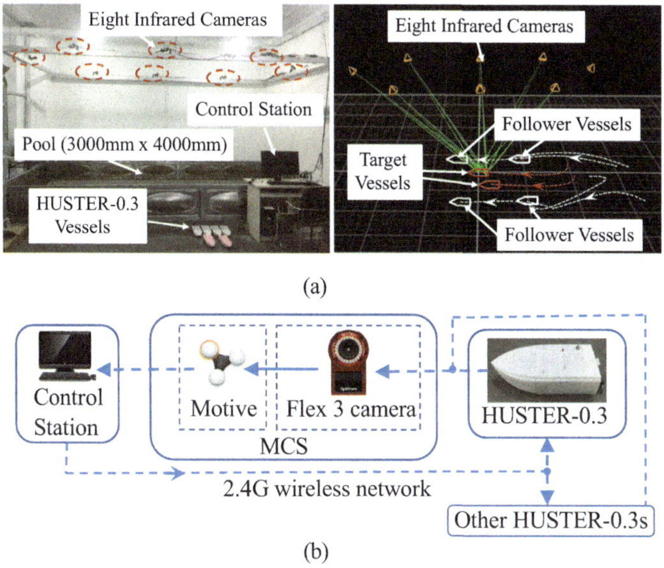

Fig. 6.2 An indoor multi-USV system for quasi-equal-distance fencing experiments. **a** Components of the indoor multi-USV system. **b** The operating procedure of the indoor multi-USV system

6.6.2 Setup and Results

In this subsection, we detail the experimental results of the fencing control with our multi-USV system. Recent advancements in USV dynamic control have been well achieved [13, 15]. A key practical approach involves reducing the USV dynamics to a second-order model and designing a two-level control system (see, for instance [15]). This methodology allowed us to implement and test the control law (6.25), which guarantees collision avoidance, using our 3D-printed HUSTER-0.3 vessels.

In the experiments, the multi-target-multi-robot system denoted as $V = I_v \cup I_t$, includes six vessels. The sensing radius R is set to 600 mm, and the safe radius r is set to 200 mm. Here, the first four vessels are designated as follower vessels (I_v), and the last two are target vessels (I_t). According to Assumption 6.1, the inter-robot communication topology is illustrated in Fig. 6.3, with targets only partially known to the follower vessels. Initially, the four follower vessels are randomly positioned in the pool, and their initial velocities are set to $v_i = 0$ mm/s for $i = 1, 2, 3, 4$. The target vessels are also placed randomly, with their velocities defined as $v_5 = [-10\sin(0.2t + 1) - 40, 0]^\top$ mm/s and $v_6 = [-10\cos(0.4t + 1) - 40, 0]^\top$ mm/s, respectively, under Assumption 6.3. The fencing radius ρ is set to 600 mm. Given these target velocities, the estimator gain β_p for the estimator (6.7) is chosen as 2.5, with $\sigma_p = 30$ mm, which ensures compliance with the inequality (6.10) as stated in Lemma 6.3. Additionally, following Theorem 6.2, the controller parameters for (6.25) are set to $k_1 = 0.5$, $k_2 = 5$, and $k_3 = 0.01$.

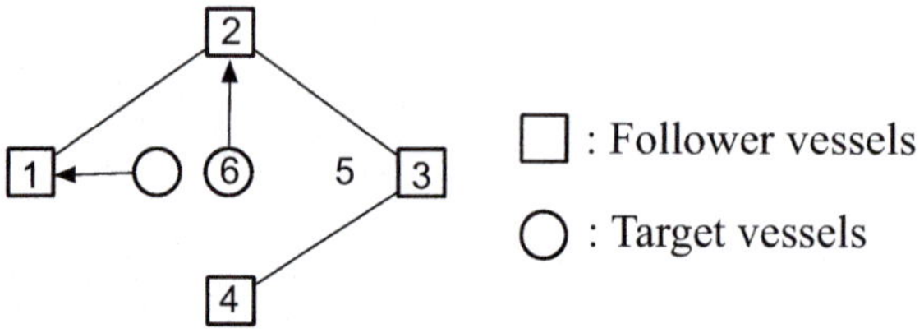

Fig. 6.3 The topology of the multi-robot-multi-target USVs

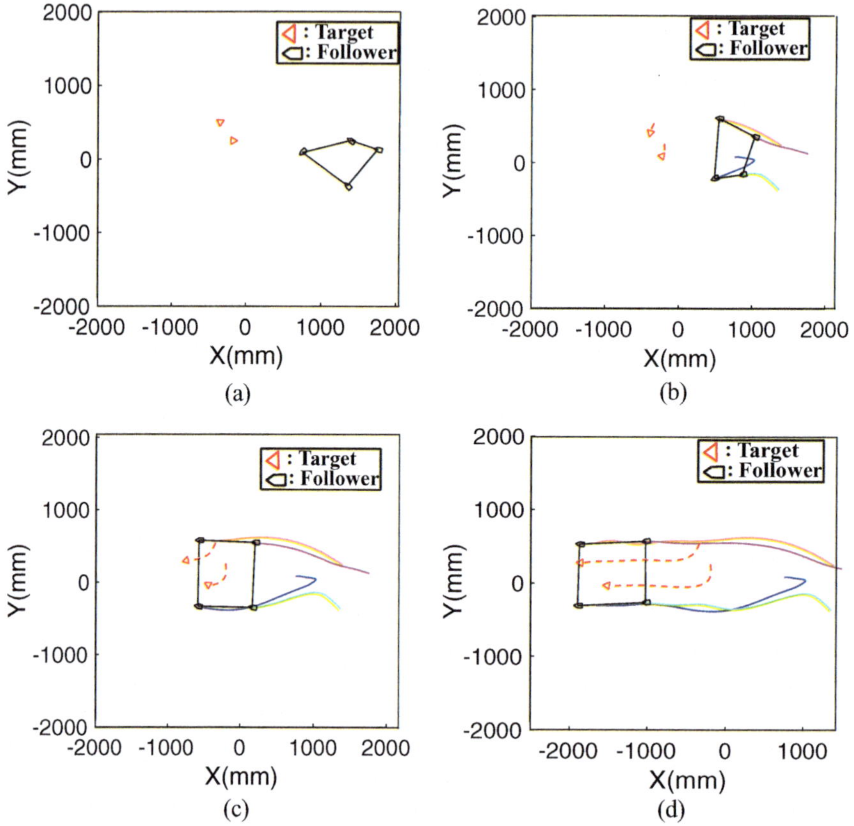

Fig. 6.4 A quasi-equal-distance fencing process of motional target USVs with the follower USVs, where the colored lines are the moving trajectories of the four vessels, while the red dashed lines are two targets. **a** Initial positions. **b** Chasing stage. **c** Catching up stage. **d** Target fencing

The results of the multi-target-multi-robot system $V = I_v \cup I_t$ are shown in Fig. 6.4, where the four follower vessels (I_v) successfully encircle the two moving targets. For a detailed view of the fencing operation, refer to the snapshot in Fig. 6.5, which corresponds to Fig. 6.4d. Figure 6.6a illustrates that the estimated errors $\widehat{x}_i$ for $i = 1, 2, 3, 4$ converge to the target center $\bar{r}$ within 5 s, adhering to

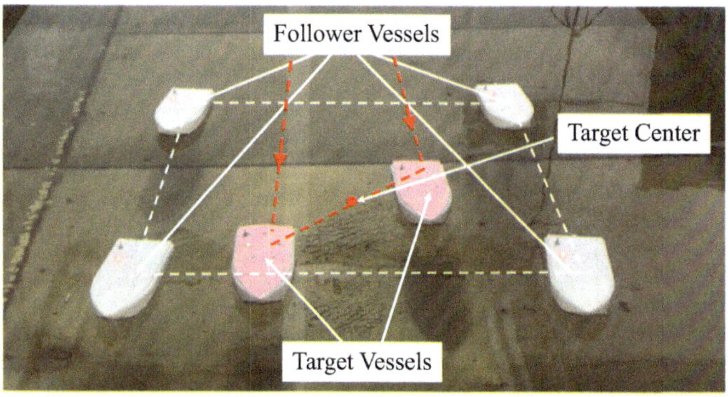

Fig. 6.5 A snapshot of the equally fencing behavior with multiple vessels, which corresponds to Fig. 6.4d. The four white vessels form a follower set and the two red vessels are the targets

the specified boundary. This confirms the effectiveness of the estimator (6.7). Similarly, Fig. 6.6b shows that the fencing errors e_i for $i = 1, 2, 3, 4$ reduce to within 200 mm in less than 20 s, validating the proposed control scheme (6.25). In Fig. 6.7a, the distances between the USVs consistently remain above 200 mm, ensuring that collisions are avoided. Additionally, Fig. 6.7b demonstrates that the phase of the inter-USV arrangement shifts toward the center of the targets and converges to 90°, which verify the practical application of Theorem 6.2.

6.7 Conclusion

In this chapter, we have developed a distributed control strategy for a nonlinear second-order MRS designed to fence a group of moving targets while ensuring collision avoidance. Our approach has demonstrated that the robots can successfully surround multiple moving targets with an even distribution around a common circle. By incorporating potential functions to manage interactions among robots and with the estimated target center, the strategy has also ensured obstacle avoidance. We have detailed the conditions for maintaining the stability of the closed-loop system and provided experimental results from our multi-HUSTER-0.3 USV system to illustrate the practical effectiveness of the proposed method, which shows significant potential for real-world applications, including marine resource management and multi-USV system operations.

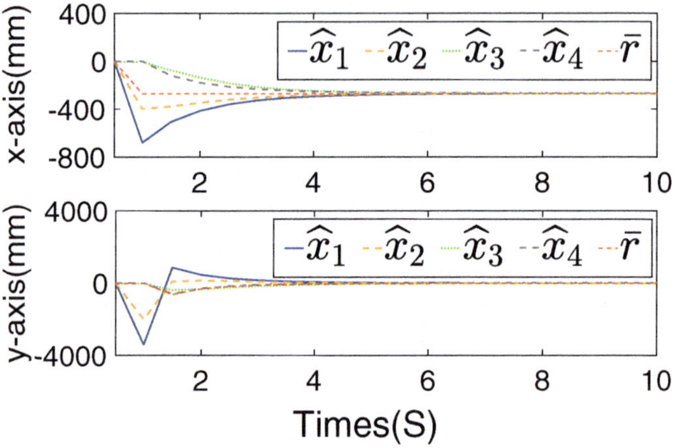

(a)

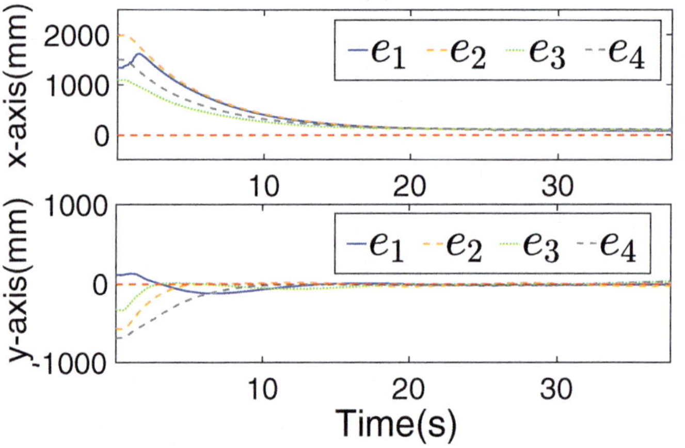

(b)

Fig. 6.6 **a** The transition of estimate $\widehat{x}_i$, $i = 1, 2, 3, 4$ of the center of target USVs with collision avoidance for follower USVs. **b** The transition of position fencing error e_i, $i = 1, 2, 3, 4$ for follower USVs

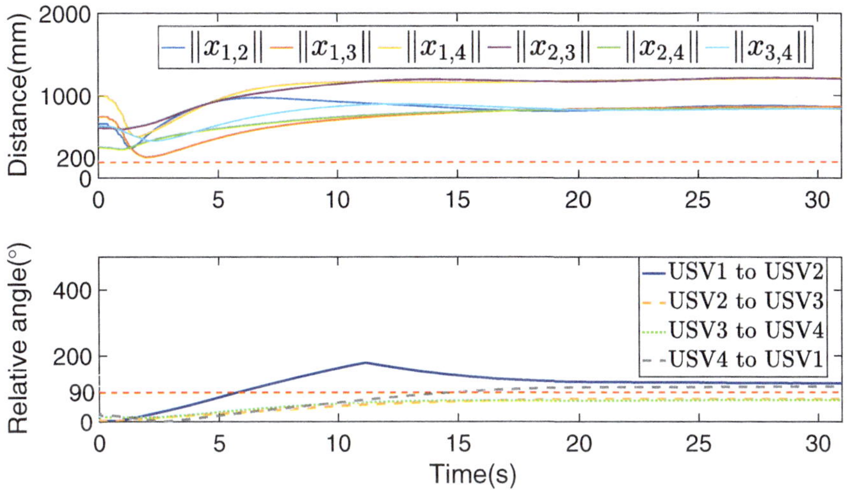

Fig. 6.7 **a** The transition of distance norm $\|x_{i,k}\|$, $i \neq k \in I_t$ between each pair of follower USVs, which verifies the collision avoidance. **b** The transition of the inter-USV phase toward the center of targets finally converges to 90°

References

1. C.W. Reynolds, Flocks, herds, and schools: a distributed behavioral model. Comput. Graph. **21**(4) (1987)
2. T. Vicsek, A. Czirók, E. Ben-Jacob, I. Cohen, O. Shochet, Novel type of phase transition in a system of self-driven particles. Phys. Rev. Lett. **75**(6), 1226–1229 (1995)
3. A. Ponomarev, Z. Chen, H.-T. Zhang, Discrete-time predictor feedback for consensus of multiagent systems with delays. IEEE Trans. Autom. Control **63**(2), 498–504 (2017)
4. H.-T. Zhang, Z. Chen, X. Mo, Effect of adding edges to consensus networks with directed acyclic graphs. IEEE Trans. Autom. Control **62**(9), 4891–4897 (2017)
5. H. Meng, H.-T. Zhang, Z. Wang, G. Chen, Event-triggered Control for Semi-global Robust Consensus of a Class of Nonlinear Uncertain Multi-agent Systems. IEEE Trans. Autom. Control **65**(4), 1683–1690 (2020)
6. H. Zhang, G. Feng, H. Yan, Q. Chen, Observer-based output feedback event-triggered control for consensus of multi-agent systems. IEEE Trans. Ind. Electron. **61**(9), 4885–4894 (2013)
7. Z. Chen, H.-T. Zhang, No-beacon collective circular motion of jointly connected multi-agents. Automatica **47**(9), 1929–1937 (2011)
8. R. Olfati-Saber, R.M. Murray, Consensus problems in networks of agents with switching topology and time-delays. IEEE Trans. Autom. Control **49**(9), 1520–1533 (2004)
9. W. Ren, R.W. Beard, Consensus seeking in multiagent systems under dynamically changing interaction topologies. IEEE Trans. Autom. Control **50**(5), 655–661 (2005)
10. Z. Chen, H.-T. Zhang, A remark on collective circular motion of heterogeneous multi-agents. Automatica **49**(5), 1236–1241 (2013)
11. Z. Peng, J. Wang, D. Wang, Containment maneuvering of marine surface vehicles with multiple parameterized paths via spatial-temporal decoupling. IEEE/ASME Trans. Mechatronics **22**(2), 1026–1036 (2016)
12. J. Mei, W. Ren, G. Ma, Distributed containment control for Lagrangian networks with parametric uncertainties under a directed graph. Automatica **48**(4), 653–659 (2012)

13. Z. Peng, J. Wang, D. Wang, Distributed containment maneuvering of multiple marine vessels via neurodynamics-based output feedback. IEEE Trans. Ind. Electron. **64**(5), 3831–3839 (2017)
14. Y. Lan, G. Yan, Z. Lin, Distributed control of cooperative target enclosing based on reachability and invariance analysis. Syst. Control Lett. **59**(7), 381–389 (2015)
15. B. Liu, Z. Chen, H.-T. Zhang, X. Wang, T. Geng, H. Su, J. Zhao, Collective dynamics and control for multiple unmanned surface vessels. IEEE Trans. Control Syst. Technol. **28**(6), 2540–2547 (2019)
16. Z. Chen, A cooperative target-fencing protocol of multiple vehicles. Automatica **107**, 591–594 (2019)
17. F. Lin, X. Dong, B.-M. Chen, K.-Y. Lum, T.H. Lee, A robust real-time embedded vision system on an unmanned rotorcraft for ground target following. IEEE Trans. Ind. Electron. **59**(2), 1038–1049 (2011)
18. J. Guo, G. Yan, Z. Lin, Local control strategy for moving-target-enclosing under dynamically changing network topology. Syst. Control Lett. **59**(10), 654–661 (2010)
19. M. Zhang, H.H.T. Liu, Game-theoretical persistent tracking of a moving target using a unicycle-type mobile vehicle. IEEE Trans. Ind. Electron. **61**(11), 6222–6233 (2014)
20. X. Yu, L. Liu, Distributed circular formation control of ring-networked nonholonomic vehicles. Automatica **68**, 92–99 (2016)
21. C. Li, L. Chen, Y. Guo, Y. Lyu, Cooperative surrounding control with collision avoidance for networked Lagrangian systems. J. Franklin Inst. **355**(12), 5182–5202 (2018)
22. F. Chen, W. Ren, Y. Cao, Surrounding control in cooperative agent networks. Syst. Control Lett. **59**(11), 704–712 (2010)
23. S. Shoja, M. Baradarannia, F. Hashemzadeh, M. Badamchizadeh, P. Bagheri, Surrounding control of nonlinear multi-agent systems with non-identical agents. ISA Trans. **70**, 219–227 (2017)
24. Y.J. Shi, R. Li, K.L. Teo, Cooperative enclosing control for multiple moving targets by a group of agents. Int. J. Control **88**(1), 80–89 (2015)
25. K.H. Khalil, *Nonlinear Systems* (Prentice Hall, Upper Saddle River, 2002)
26. Y. Jiang, Z. Peng, D. Wang, C.L.P. Chen, Line-of-sight target enclosing of an underactuated autonomous surface vehicle with experiment results. IEEE Trans. Ind. Electron. **16**(2), 832–841 (2020)
27. R. Li, Y. Shi, Y. Song, Multi-group coordination control of multi-agent system based on smoothing estimator. IET Control Theory Appl. **10**(11), 1224–1230 (2016)

Chapter 7
Coordinated Multiple-Scattered-Targets Fencing via Flexible Subgrouping and Ordering

7.1 Introduction

In the previous Chap. 6, we studied the multi-target fencing via adaptive controller, where the dynamics of the targets are always limited to cluster together. Such kind of scenario inevitably limits the application of the designed controller. A straightforward idea is to form a big circle of robots to fence the small circle formed by the target. However, there still exist some challenging scenarios where the targets may be scattered at different positions, which motivates the fencing controller design via flexible subgrouping and ordering in this chapter. In what follows, we will introduce the related background.

In recent years, substantial progress has been made in multi-robot research across various domains, such as robotics, control systems, and autonomous driving [1–4], largely due to its inherent efficiency and wide-ranging applications. Coordinated target fencing, a critical problem within multi-robot research, involves guiding a group of robots to form one or more convex hulls to protect a target(s) [5], and it has gained significant attention. Approaches to target fencing can be broadly classified into two categories—fixed-ordering and flexible-ordering—depending on whether the convex hull is formed with a predetermined spatial sequence or not.

In fixed-ordering strategies, robots are controlled to reach designated relative positions within the convex hull, resulting in a fixed spatial sequence among the robots. Such kind of method is simple and suitable for many practical applications. Early studies concentrated on target fencing for stationary targets [5, 6], and these were later extended to address moving targets [7] under varying network topologies. However, these approaches [5–7] primarily deal with single-target scenarios, leaving the more complex cases involving multiple targets largely unexplored, mainly due to the unpredictable and random nature of target motion. To tackle this challenge, some works have confined multiple targets within a bounded area and then enclosed them within a larger convex hull using estimators for the targets' centers [8–11], reducing the need for complex dynamic tracking. For more difficult cases where multiple targets are dispersed across different regions, robots are typically grouped in advance

B.-B. Hu and H.-T. Zhang, *Multi-Robot Systems*, Cognitive Intelligence and Robotics,
https://doi.org/10.1007/978-981-97-9214-6_7

and then tasked with fencing each target individually using fixed-ordering methods [12, 13]. Nevertheless, these fixed-ordering approaches [5–13] predetermine the robots' desired positions and orderings, limiting adaptability and potentially increasing the robots' travel distances. Additionally, the pre-assignment of robot subgroups in [12, 13] further restricts flexibility, as it locks in the number and indices of robots, making it difficult to adjust to dynamic environments.

Meanwhile, a different line of research has focused on more efficient flexible-ordering strategies for target fencing, allowing robots to form a convex hull without enforcing a fixed spatial sequence [14]. For instance, a cooperative controller was proposed in [15] to fence a stationary target, while an output-regulation-based controller was designed in [16] for fencing a target using unmanned surface vessels. This approach was later adapted to constant-velocity targets [17]. For more challenging cases involving targets with varying velocities, a distributed control law was introduced in [18] to achieve circular fencing. More recently, a cooperative label-free controller using a dynamic regulator was developed in [19] for fencing targets with periodic dynamics. However, most existing flexible-ordering approaches [15–19] focus exclusively on single-target scenarios and have yet to address the critical yet complex problem of multiple scattered targets (MSTs). While some studies [20, 21] explored multi-robot allocation strategies that could potentially be integrated with existing flexible fencing techniques [15–19] to divide robots into subgroups, this two-step allocation-control design treats each stage separately, leading to high computational complexity and limiting the practical application of these methods.

Drawing inspiration from minimum-energy task execution [22] and the flexible-ordering fencing approach for a single target [19], we develop an FSOC algorithm to manage a team of robots tasked with fencing multiple scattered targets (MSTs) of varying speeds within distinct convex hulls formed by different robot subgroups in dynamic environments. The proposed FSOC method encodes the three key subtasks—flexible subgrouping, target convergence, and neighboring collision avoidance–into CBF constraints, framing the problem as a decentralized constraint-based task-execution optimization. By incorporating fencing matching conditions into both hard integer constraints and soft cost functions, the algorithm allows for adaptive subgrouping with flexible sizes and member assignments. Within each subgroup, the interplay between target-convergence constraints and neighboring collision-avoidance constraints results in the formation of a convex hull with an arbitrary spatial ordering for each target. The main contributions of this chapter can be summarized as follows.

1. Design the FSOC algorithm to guide a team of robots in dynamically dividing into subgroups with flexible sizes and compositions to handle multiple scattered targets. Each target is then enclosed within a rigid convex hull formed by its corresponding subgroup, allowing for flexible spatial orderings among the robots within the formation.
2. Rigorously ensure that each target's fencing formation converges asymptotically, despite the strong nonlinear interactions resulting from flexible subgrouping and spatial orderings.

3. Optimize the process by simultaneously computing both the optimal subgrouping vector and control inputs, which reduces computational costs and eliminates the need for separate planning steps.
4. Conduct comprehensive 2D and 3D simulations to verify the proposed algorithm's flexibility across different starting configurations, which includes its robustness to sudden robot breakdown and unexpected target appearances, its adaptability in obstructed environments, and its scalability for complex high-dimensional fencing tasks.

The remainder of this chapter is structured as follows: Sect. 7.2 provides the necessary preliminaries. Section 7.3 formulates the problem. Section 7.4 details the FSOC optimization process. Section 7.5 presents the convergence analysis. Section 7.6 discusses the simulation results. Finally, Sect. 7.7 concludes the chapter.

7.2 Preliminaries

In this section, we will introduce the necessary ingredients for MST convoying missions.

7.2.1 *Task Set*

We consider a desired task set $\mathcal{T} \subseteq \mathbb{R}^n$ to be

$$\mathcal{T} := \{\xi \in \mathbb{R}^n \mid \phi(\xi) \leq 0\}, \tag{7.1}$$

where $\xi := [\xi_1, \ldots, \xi_n]^\top$ denotes the dummy coordinates in the n-dimensional Euclidean space, and $\phi : \mathbb{R}^n \to \mathbb{R}$ is a twice continuously differentiable function [22]. By substituting a specific point $p_0 \in \mathbb{R}^n$ into $\phi(\xi)$, the resulting value $\phi(p_0)$ provides a rough approximation of the distance $\mathrm{dist}(p_0, \mathcal{T})$ between the point p_0 and the set $\mathcal{T}$. Consequently, $\phi(p_0)$ can be used to represent the task-convergence errors at p_0 when $\phi(p_0) > 0$.

Assumption 7.1 ([23]) For any $\kappa \in \mathbb{R}^+$ and for any point p_0 located outside the set $\mathcal{T}$ as defined in Eq. (7.1), it holds that $\inf \phi(p_0) : \mathrm{dist}(p_0, \mathcal{T}) \geq \kappa > 0$.

Assumption 7.1 rigorously ensures that $\lim_{t\to\infty} \mathrm{dist}(p_0(t), \mathcal{T}) = 0$ if $\lim_{t\to\infty} \phi(p_0(t)) \leq 0$.

7.2.2 Multi-robot Task Execution

The initial component of the MST fencing mission is a multi-robot system consisting of N robots, with indices in the set $\mathcal{V} = \{1, \ldots, N\}$. Each robot operates according to single-integrator dynamics as outlined in [23], represented by the following equation,

$$\dot{x}_i = u_i, i \in \mathcal{V}. \tag{7.2}$$

Here, $x_i \in \mathbb{R}^n$ and $u_i \in \mathbb{R}^n$ denote the position and control input of robot i, respectively, where $i \in \mathcal{V}$. The complexity of the MST fencing primarily resides in the coordination among multiple robots, rather than in the intricate dynamic control. For simplicity, we use a basic single-integrator model as specified in Eq. (7.2). In (7.2), the input u_i can be viewed as the upper-level velocity command assigned to each robot, applicable when interacting with robots that have distinct dynamics [23]. For further details, see Remark 7.3.

Definition 7.1 (*Minimum-energy task execution* [22]) To achieve minimum-energy task execution for a desired task set $\mathcal{T}$ as specified in (7.1), the multi-robot system $\mathcal{V}$ governed by (7.2) solves the following constraint-based optimization problem

$$\min_{u_i, \delta_i} \; \|u_i\|^2 + |\delta_i|^2 \tag{7.3a}$$

$$\text{s.t. } \frac{\partial h(x_i)}{\partial x_i^{\mathsf{T}}} u_i \geq -\gamma\big(h(x_i)\big) - \delta_i, i \in \mathbb{Z}_1^N, \tag{7.3b}$$

where x_i and u_i are defined in Eq. (7.2). The function $h(x_i) = -\phi(x_i)$, which adheres to constraint (7.3b), represents a CBF [24]. The slack variable $\delta_i \in \mathbb{R}^+$ measures the extent of task violation when $x_i \notin \mathcal{T}$. Moreover, $\gamma(\cdot) : (-b, a) \to \mathbb{R}$ is an extended class $\mathcal{K}$ function with $a, b \in \mathbb{R}^+$, which is strictly increasing and satisfies $\gamma(0) = 0$ [24].

Definition 7.1 outlines a core framework for encoding both flexible subgrouping and flexible-ordering fencing, which will be detailed further in Definitions 7.2 and 7.3.

Remark 7.1 According to the Karush–Kuhn–Tucker (KKT) conditions [26] for the minimum-energy task execution problem in (7.3), [22] demonstrates that the optimal solutions for u_i^*, δ_i^*, along with the optimal Lagrange multiplier $\lambda^* \geq 0$, are determined by the constraint $g_i := -(\partial h(x_i)/\partial x_i^{\mathsf{T}})u_i - \gamma\big(h(x_i)\big) - \delta_i$ in (7.3b).

$$u_i^* = \frac{-\gamma\big(h(x_i)\big)}{1 + \|\frac{\partial h(x_i)}{\partial x_i^{\mathsf{T}}}\|^2} \frac{\partial h(x_i)}{\partial x_i}, \; \lambda^* = \frac{-2\gamma\big(h(x_i)\big)}{1 + \|\frac{\partial h(x_i)}{\partial x_i^{\mathsf{T}}}\|^2}. \tag{7.4}$$

Due to the properties of the extended class $\mathcal{K}$ function $\gamma(\cdot)$ in Eq. (7.3), it follows from Eq. (7.4) that

$$
\begin{aligned}
&(1)\ \lambda^* > 0 \iff h(x_i) < 0 \iff x_i \notin \mathcal{T}, \\
&(2)\ \lambda^* = 0 \iff h(x_i) = 0 \implies x_i \in \mathcal{T}.
\end{aligned} \tag{7.5}
$$

According to the complementary slackness condition $\lambda^* g_i = 0$, if $\lambda^* > 0$, then the boundary of the constraint (7.3b) must be active, which implies $g_i = 0$. Furthermore, given that $h(x_i) = -\phi(x_i)$ and $\mathcal{T} := \{\sigma \in \mathbb{R}^n \mid \phi(\sigma) \leq 0\}$ in Eq. (7.1), it follows that

$$
\begin{aligned}
\lambda^* > 0 &\iff x_i(t) \notin \mathcal{T} \implies g_i = 0, \\
\lambda^* = 0 &\implies x_i(t) \in \mathcal{T} \implies g_i \leq 0.
\end{aligned} \tag{7.6}
$$

The connection between λ^* and $x_i(t) \in \mathcal{T}_i$, as outlined in Eq. (7.6), will be used in proving Lemmas 7.1 and 7.2 later in this chapter.

7.2.3 Multiple Scattered Targets

The second component of the MST fencing mission consists of M scattered targets, indexed by the set $\mathcal{M} = 1, \ldots, M$. Each target moves with a distinct velocity as described below

$$
\dot{r}_k = v_k, k \in \mathcal{M}, \tag{7.7}
$$

where $r_k \in \mathbb{R}^n$ and $v_k \in \mathbb{R}^n$ represent the position and velocity of target k, respectively. It is assumed that v_k is continuously differentiable.

Assumption 7.2 For any pair of distinct targets, the distance between them satisfies $|r_k(t) - r_l(t)| \gg R$ for all $k \neq l \in \mathcal{M}$ and for all $t > 0$, where $R \in \mathbb{R}^+$ denotes the sensing radius.

Assumption 7.2 ensures that each scattered target $k \in \mathcal{M}$ is independently fenced by different subgroups. However, this assumption has a limitation: in practical scenarios, the distance between any two scattered targets may not always be adequately maintained. When targets are close to each other, collision avoidance measures for robots can impact the convex-hull fencing.

Assumption 7.3 The number of targets M and the number of robots N satisfy the condition $M(n+1) \leq N$.

Assumption 7.3 is valid because $n+1$ robots are required to form a convex hull around each target in an n-dimensional Euclidean space. If there are fewer than $n+1$ robots, some targets might not be properly enclosed within the convex hull. For example, in a 2D space, $n = 2$ robots can only create a line, and in a 3D space, $n = 3$ robots can only form a plane, not a full convex body. The limitation arises if $M(n+1) > N$, which means there may be insufficient robots to effectively fence all the scattered targets in practice.

7.3 Problem Formulation of Fencing Multiple Scattered Targets

7.3.1 *Flexible Subgrouping for MSTs*

We start by defining a subgrouping vector α_i for robot i as $\alpha_i := [\alpha_{i,1}, \ldots, \alpha_{i,M}]^\top \in \mathbb{R}^M$ [25] where each entry $\alpha_{i,k}, k \in \mathcal{M}$ satisfies

$$\alpha_{i,k} = \begin{cases} 1, & \text{if robot } i \text{ is assigned to fence target } k, \\ 0, & \text{otherwise.} \end{cases} \tag{7.8}$$

Then, we can define M subsets $\mathbb{V}_k \subset \mathcal{V}$ for the corresponding M targets to be

$$\mathbb{V}_k := \{i \in \mathcal{V} \mid \alpha_{i,k} = 1\}, \tag{7.9}$$

and $\boldsymbol{\alpha} := [\alpha_1, \ldots, \alpha_N]^\top \in \mathbb{R}^{N\times M}$ to be the subgrouping matrix for the robot group $\mathcal{V}$, respectively.

Definition 7.2 (*Multi-robot flexible subgrouping*) All robots $\mathcal{V}$ in Eq. (7.2) are flexibly divided into M subgroups $\mathbb{V}_k$ as described in (7.9) for M scattered targets if the following two mild conditions are met,

1. Each robot fences only one target, i.e., $\boldsymbol{\alpha}\mathbf{1}_M = \mathbf{1}_N$.
2. Each target is fenced by at least $n+1$ robots in the n-dimensional Euclidean space, i.e., $\mathbf{1}_N^\top\boldsymbol{\alpha} \geq (\mathbf{n}+\mathbf{1})_M^\top$ with $(\mathbf{n}+\mathbf{1})_M := [n+1, \ldots, n+1]^\top \in \mathbb{R}^M$.

In Definition 7.2, Condition (1) ensures that each robot is assigned to only one subset $\mathbb{V}_k$ in Eq. (7.9), meaning that $\cup_{k=1}^M \mathbb{V}_k = \mathcal{V},\ \mathbb{V}_k \cap \mathbb{V}_l = \emptyset, \forall k \neq l \in \mathcal{M}$. Condition (2) specifies only the minimum number of robots needed to form a convex hull, without dictating the exact number or indexing of robots, i.e., $|\mathbb{V}_k| \geq n+1$ for all $k \in \mathcal{M}$. Combining Conditions (1) and (2) allows Definition 7.2 to be flexible for the number and indices of subgroups. Examples illustrating this flexibility are provided in Fig. 7.1.

7.3.2 *Flexible-Ordering Fencing for Each Specific Target*

Next, we will introduce flexible-ordering fencing, which adds further flexibility to the algorithm by allowing for varied spatial ordering sequences.

Let $\mathrm{co}(\mathbb{V}_k)$ denote the convex hull formed by the robot subgroup $\mathbb{V}_k$ [15], defined as $\mathrm{co}(\mathbb{V}_k) := \{\sum_{i\in\mathbb{V}_k} \lambda_i x_i \mid \lambda_i \geq 0, i \in \mathbb{V}_k, \text{and} \sum_{i\in\mathbb{V}_k} \lambda_i = 1\}$, where x_i represents the position of robot $i \in \mathbb{V}_k$. The distance from a specific target k to the convex hull $\mathrm{co}(\mathbb{V}_k)$ is given by: $\mathrm{dist}\big(r_k, \mathrm{co}(\mathbb{V}_k)\big) = \inf\{\|r_k - s\| \mid s \in \mathrm{co}(\mathbb{V}_k)\}$.

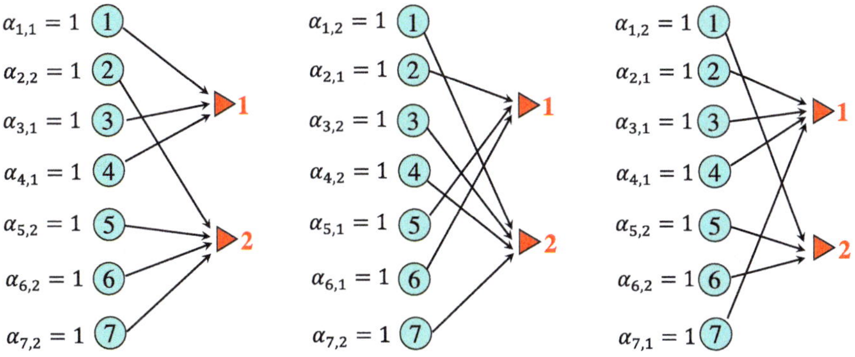

(a) Allocation 1 (b) Allocation 2 (c) Allocation 3

Fig. 7.1 Three examples of the flexible subgrouping for a seven-robot two-target system in Definition 7.2, where the blue circles, red triangles, and black arrows denote the robots, targets, and the robots assigned to the targets, respectively. **a** Subgroups 1: $\mathbb{V}_1 = \{1, 3, 4\}$ and $\mathbb{V}_2 = \{2, 5, 6, 7\}$. **b** Subgroups 2: $\mathbb{V}_1 = \{2, 5, 6\}$ and $\mathbb{V}_2 = \{1, 3, 4, 7\}$. **c** Subgroups 3: $\mathbb{V}_1 = \{2, 3, 4, 7\}$ and $\mathbb{V}_2 = \{1, 5, 6\}$

Definition 7.3 (*Flexible-ordering rigid-formation fencing* [19]) For each target $k \in \mathcal{M}$, governed by Eq. (7.7), the robot subgroup $\mathbb{V}_k$, following Eqs. (7.2) and (7.9), will asymptotically establish flexible-ordering rigid-formation fencing if the following three conditions are satisfied

1. (**Flexible-ordering fencing**) The robots in subset $\mathbb{V}_k$ fence the scattered target k within the interior of $\text{co}(\mathbb{V}_k)$, using an arbitrary ordering sequence, i.e., $\lim_{t\to\infty} \text{dist}\big(r_k(t), \text{co}(\mathbb{V}_k(t))\big) = 0$ with r_k given in (7.7).
2. (**Rigid-formation maneuvering**) The robots in subset $\mathbb{V}_k$ establish and maintain a rigid formation around the moving target k, i.e., $\lim_{t\to\infty}\{u_i(t) - v_k(t)\} = \mathbf{0}_n, i \in \mathbb{V}_k$, where v_k is given in (7.7).
3. (**Collision avoidance**) The robots in subset $\mathbb{V}_k$ continuously avoid undesired inter-robot collisions, i.e., $\|x_i(t) - x_j(t)\| > r, \forall t > 0, i \neq j \in \mathbb{V}_k$, where $r \in \mathbb{R}^+$ is a collision radius.

In Definition 7.3, Objective (1) does not impose a predetermined spatial ordering for the robots in the convex hull $\text{co}(\mathbb{V}_k)$. Consequently, $\text{co}(\mathbb{V}_k)$ can be formed with robots arranged in any flexible ordering.

Problem 7.1 To simultaneously determine the optimal subgrouping vector and control inputs α_i^*, u_i^* for each robot i governed by Eq. (7.2), such that flexible subgrouping and ordering fencing for the scattered targets (MSTs) governed by Eq. (7.7) is achieved, satisfying Conditions (1)–(2) in Definition 7.2 and Objectives (1)–(3) in Definition 7.3.

7.4 FSOC Optimization

To solve Problem 7.1, we begin by encoding the subtasks related to target convergence and neighboring collision avoidance. For robots $i \in \mathbb{V}k$ assigned to target $k \in \mathcal{M}$, we introduce the desired target-convergence task set $\mathcal{T}^r_{i,k}$, analogous to the set $\mathcal{T}$ in Eq. (7.1), as follows:

$$\mathcal{T}^r_{i,k} = \left\{\xi_i := [\xi_{i,1}, \ldots, \xi_{i,n}]^\mathrm{T} \in \mathbb{R}^n \;\middle|\; \phi^r_{i,k}(\xi_i) := \|\xi_i - r_k\| \leq 0\right\}, i \in \mathbb{V}_k, k \in \mathcal{M}, \tag{7.10}$$

where ξ_i is the ith dummy coordinates and r_k is the position given in Eq. (7.7). Combining with x_i in Eqs. (7.2) and (7.10) together, one has that the target-convergence error becomes

$$\phi^r_{i,k}(x_i) = \|x_i - r_k\|, i \in \mathbb{V}_k, k \in \mathcal{M}. \tag{7.11}$$

Based on the subgrouping entry $\alpha_{i,k}$ in Definition 7.2 and the CBF constraint from Definition 7.1, the target-convergence CBF constraints for $\mathcal{T}^r_{i,k}$ are given as follows (Fig. 7.2):

$$\frac{\partial \phi^r_{i,k}(x_i)}{\partial x_i^\mathrm{T}}(u_i - v_k) \leq -\gamma\left(\phi^r_{i,k}(x_i)\right) + \delta^r_{i,k} + (1 - \alpha_{i,k})\eta, \tag{7.12}$$

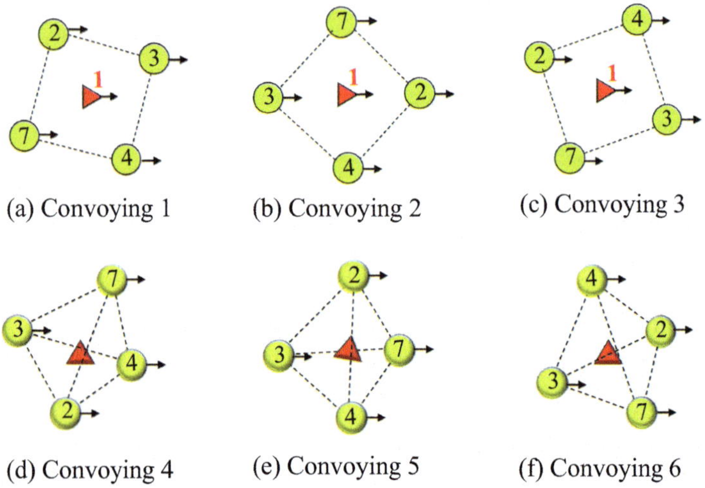

Fig. 7.2 The same-shape fencing formations for the specific target 1 and $\mathbb{V}_1 = \{2, 3, 4, 7\}$ with three different ordering sequences in Fig. 7.1c. Here, **a–c** and **d–f** are the fencing formations in 2D and 3D, respectively

where $\delta^r_{i,k} \in \mathbb{R}^+$ is a slack variable that accounts for cases where $x_i \neq r_k$, since the set $\mathcal{T}^r_{i,k}$ contains only the single point r_k. The term $\eta \in \mathbb{R}^+ \gg \delta^r_{i,k}$ serves as a penalty for $\alpha_{i,k} = 0$, and $\gamma(\cdot)$ represents the extended class $\mathcal{K}$ functions defined in Definition 7.1.

Based on the different values of $\alpha_{i,k}$, we can categorize the constraints in Eq. (7.12) into two distinct cases: ($i \in \mathbb{V}_k$ for $\alpha_{i,k} = 1$) and ($i \in \mathcal{V}\backslash\mathbb{V}_k = \{i \in \mathcal{V} | i \notin \mathbb{V}_k\}$ for $\alpha_{i,k} = 0$). Specifically, when robot i satisfies $i \in \mathbb{V}k$, the constraint in Eq. (7.12) is strictly enforced by a positive variable $\delta^r_{i,k}$ as follows:

$$\frac{\partial \phi^r_{i,k}(x_i)}{\partial x_i^{\mathsf{T}}}(u_i - v_k) \leq -\gamma\left(\phi^r_{i,k}(x_i)\right) + \delta^r_{i,k}, i \in \mathbb{V}_k. \tag{7.13}$$

Here, for robot i, the remaining $|\mathcal{V}\backslash\mathbb{V}k|$ constraints in Eq. (7.12) are governed by a higher slack value of $\delta i, k^r + \eta$, which can be expressed by

$$\frac{\partial \phi^r_{i,k}(x_i)}{\partial x_i^{\mathsf{T}}}(u_i - v_k) \leq -\gamma\left(\phi^r_{i,k}(x_i)\right) + \delta^r_{i,k} + \eta, i \in \mathcal{V}\backslash\mathbb{V}_k.$$

Let $\mathcal{N}_i$ denote the set of sensing neighbors defined as follows:

$$\mathcal{N}_i(t) = \{j \neq i \in \mathcal{V} \mid \|x_i(t) - x_j(t)\| \leq R\} \tag{7.14}$$

where $R \in \mathbb{R}^+ \in (r, \infty]$ is defined in Assumption 7.2, we define the collision-avoidance task set $\mathcal{T}^c_{i,j}$ for the neighboring robots $j \in \mathcal{N}_i$ as follows:

$$\mathcal{T}^c_{i,j} = \left\{\xi_i := [\xi_{i,1}, \ldots, \xi_{i,n}]^{\mathsf{T}} \in \mathbb{R}^n \mid \phi^c_{i,k}(\xi_i) := \frac{1}{\|\xi_i - x_j\| - r} - \frac{1}{\rho - r} \leq 0\right\}, i \in \mathcal{V}, \tag{7.15}$$

where $\rho \in \mathbb{R}^+$ satisfies $\rho \in (r, R]$ with R, r given below (7.14). Analogously, substituting x_i for ξ_i in Eq. (7.15) yields the neighboring collision-avoidance error

$$\phi^c_{i,j}(x_i) = \frac{1}{\|x_i - x_j\| - r} - \frac{1}{\rho - r}, j \in \mathcal{N}_i, \tag{7.16}$$

which implies that

$$\phi^c_{i,j}(x_i) \leq 0, \text{ if } \|x_i - x_j\| \geq \rho, \; \phi^c_{i,j}(x_i) \to \infty, \text{ if } \|x_i - x_j\| \to r. \tag{7.17}$$

Next, we can define the corresponding collision-avoidance CBF constraint for $\mathcal{T}^c_{i,j}$ to be

$$\frac{\partial \phi^c_{i,j}(x_i)}{\partial x_i^{\mathsf{T}}}(u_i - v_k) \leq -\gamma\left(\phi^c_{i,j}(x_i)\right) + \delta^c_{i,j}, j \in \mathcal{N}_i, \tag{7.18}$$

where $\delta^c_{i,j} \in \mathbb{R}^+$ is a slack variable to align with the constraint in Eq. (7.13). Moreover, it follows from the definition of $\mathcal{N}_i$ in Eq. (7.14) that the collision avoidance is naturally satisfied for robot $j \notin \mathcal{N}_i$. Based on Definitions 7.1–7.3 and the constraints given in Eqs. (7.12) and (7.18), the MST fencing problem is framed as a decentralized optimization problem with constraints for each robot i.

$$\min_{u_i, \delta^r_{i,k}, \delta^c_{i,j}, \alpha_i} \left\{ b\| 1_N^{\mathsf{T}} \boldsymbol{\alpha} - (\mathbf{n}+\mathbf{1})_M^{\mathsf{T}} \|^2 + c\Big(\sum_{k\in\mathcal{M}} (\delta^r_{i,k})^2 + \sum_{j\in\mathcal{N}_i} (\delta^c_{i,j})^2 \Big) + \sum_{k\in\mathcal{M}} \alpha_{i,k} \|u_i - v_k\|^2 \right\} \tag{7.19a}$$

$$\text{s.t. } \mathbf{C1}: \frac{\partial \phi^r_{i,k}(x_i)}{\partial x_i^{\mathsf{T}}} (u_i - v_k) + \gamma\big(\phi^r_{i,k}(x_i)\big) - (1-\alpha_{i,k})\varpi - \delta^r_{i,k} \le 0,\ k \in \mathbb{Z}_1^M, \tag{7.19b}$$

$$\mathbf{C2}: \frac{\partial \phi^c_{i,j}(x_i)}{\partial x_i^{\mathsf{T}}} (u_i - v_k) + \gamma\big(\phi^c_{i,j}(x_i)\big) - \delta^c_{i,j} \le 0,\ j \in \mathcal{N}_i(t), \tag{7.19c}$$

$$\mathbf{C3}: \alpha_i^{\mathsf{T}} \mathbf{1}_N = 1,\ \alpha_{i,k} \in \{0, 1\} \tag{7.19d}$$

where $b, c \in \mathbb{R}^+$ represent the weights of different cost terms, the cost function in Eq. (7.19a) comprises three components. The first term, $\| 1_N^{\mathsf{T}} \boldsymbol{\alpha} - (\mathbf{n}+\mathbf{1})_M^{\mathsf{T}} \|^2$, is designed to satisfy Condition 2 in Definition 7.2. The second term, $\sum_{k\in\mathcal{M}} (\delta^r_{i,k})^2 + \sum_{j\in\mathcal{N}_i} (\delta^c_{i,j})^2$, aims to construct a flexible-ordering convex hull for the assigned target k. The third term, $\alpha_{i,k}\|u_i - v_k\|^2$, focuses on maneuvering toward the assigned target k. Regarding the constraints, details of **C1** and **C2** in Eqs. (7.19b) and (7.19c) are outlined in Eqs. (7.12) and (7.18), respectively. Constraint **C3** in Eq. (7.19d) aligns with Condition 1 in Definition 7.2. Note that the subgrouping vector $\alpha_i = [\alpha_{i,1}, \ldots, \alpha_{i,M}]^T$ in Eq. (7.19a) ensures that robot i is assigned to target k since $\alpha_i^T \mathbf{1}_N = 1$ in **C3**. Meanwhile, the control input u_i in Eq. (7.19a) is responsible for the robot's movement. By incorporating α_i into the CBF constraints (Eq. (18b)), u_i can be regulated to drive robot i toward target k.

Assumption 7.4 The initial positions of any two robots and targets are such that $\|x_i(0) - x_j(0)\| > r$, $\|x_i(0) - r_k(0)\| > 0$ for all $i \neq j \in \mathcal{V}$ and $k \in \mathcal{M}$.

Assumption 7.5 The cost weights b, c in Eq. (7.19) satisfy $b \gg c$.

Assumption 7.5 implies $\| 1_N^{\mathsf{T}} \boldsymbol{\alpha} - (\mathbf{n}+\mathbf{1})_M^{\mathsf{T}} \|$ dominates in the cost function (7.19a), which is necessary for Assumption 7.6 later.

Remark 7.2 Unlike traditional collision-avoidance constraints found in previous works [22, 24], which do not incorporate slack variables and achieve collision avoidance through forward invariance, the constraints **C2** in Eq. (7.19c) introduce an additional slack variable $\delta^c_{i,j}$. This adjustment helps align with the constraint **C1** and ensures strong duality and the existence of a solution that meets the KKT

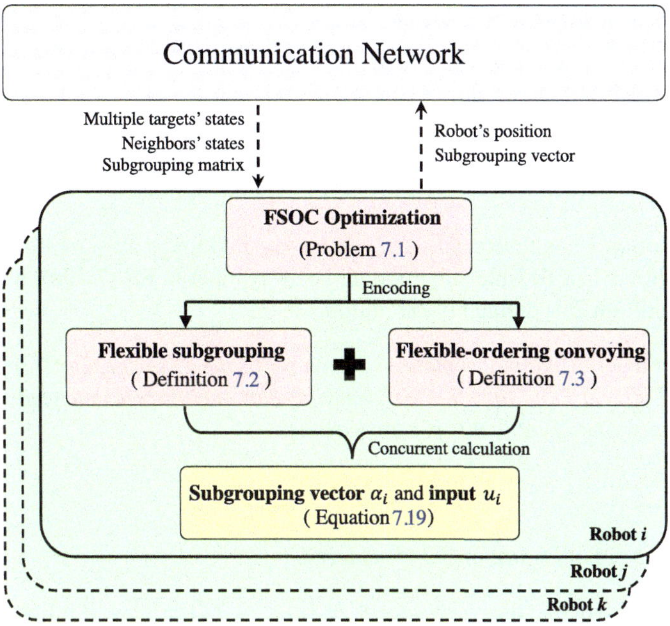

Fig. 7.3 Structure illustration of the FSOC optimization in Eq. (7.19)

conditions. Consequently, by using such slack variables, we will later demonstrate that flexible-ordering fencing can be proven in Lemma 7.2. However, introducing these slack variables $\delta_{i,j}^c$ may lead to issues where inter-robot distances might not fully satisfy the desired collision-avoidance requirements as defined by the task set $\mathcal{T}_{i,j}^c$ (i.e., $\|x_i - x_j\| > \rho \Leftrightarrow x_i \notin \mathcal{T}_{i,j}^c$). As a result, the traditional forward invariance approach might not be sufficient to prove collision avoidance. Nevertheless, we will show later that collision avoidance is still ensured through a new proof technique. Specifically, we use the boundedness of a Lyapunov function to demonstrate that collisions among robots will not occur. For technical details, refer to the proof of Lemma 7.1 (Fig. 7.3).

7.5 Convergence Analysis

Due to the presence of mixed-integer constraints in Eqs. (7.19b) and (7.19d), which complicate the direct analysis, we will break down the main results into two steps to facilitate understanding for the readers. In Step 1, we prove that the robots $\mathcal{V}$ will be flexibly and stably separated into M subgroups for M scattered targets, i.e., Conditions (1)–(2) in Definition 7.2. In Step 2, we prove that each target will be fenced by a flexible-ordering convex hull, i.e., Objectives (1)–(3) in Definition 7.3.

7.5.1 Subgrouping Convergence

Assumption 7.6 The cost term $\|1_N^\top \boldsymbol{\alpha} - (\mathbf{n}+\mathbf{1})_M^\top\|$ in Eq. (7.19) will ultimately converge to its minimum value and remain constant after a finite time $T > 0$. Specifically, we have $\lim_{t\to T} \|1_N^\top \boldsymbol{\alpha}(t) - (\mathbf{n}+\mathbf{1})_M^\top\| = \|1_N^\top \boldsymbol{\alpha}^* - (\mathbf{n}+\mathbf{1})_M^\top\|$, and $\|1_N^\top \boldsymbol{\alpha}(t) - (\mathbf{n}+\mathbf{1})_M^\top\| = \|1_N^\top \boldsymbol{\alpha}^* - (\mathbf{n}+\mathbf{1})_M^\top\|, \forall t > T$, with the optimal $\boldsymbol{\alpha}^*$.

Assumption 7.6 is reasonable and can be satisfied under Assumption 7.5. Given that Condition 1 of flexible subgrouping is always met in Eq. (7.19d), it suffices to prove Condition 2 as defined in Definition 7.2.

Proposition 7.1 *Under Assumptions 7.3 and 7.6, the robot group $\mathcal{V}$ will become separated after a finite time $T > 0$, ensuring that each target is fenced by at least $n+1$ robots. Specifically, this is expressed as*

$$\mathbf{1}_N^\top \boldsymbol{\alpha}(t) \geq (\mathbf{n}+\mathbf{1})_M^\top, \forall t \geq T, \tag{7.20}$$

in Condition (2) of Definition 7.2 is achieved.

Proof First, we define the row vector $\boldsymbol{\zeta} := [\zeta_1, \ldots, \zeta_m] = \mathbf{1}_N^\top \boldsymbol{\alpha}$. It follows that $\zeta_k \geq n+1$ for all $k \in \mathcal{M}$, which is equivalent to $\mathbf{1}_N^\top \boldsymbol{\alpha} \geq (\mathbf{n}+\mathbf{1})_M^\top \iff \zeta_k \geq n+1, \forall k \in \mathcal{M}$.

We will prove by contradiction. Without loss of generality, assume that there exists at least one target l such that, for $t \geq T$, it is fenced by fewer than $n+1$ robots, i.e.,

$$|\mathbb{V}_l(t)| = \zeta_l(t) < n+1, \forall t \geq T. \tag{7.21}$$

Then, the contradiction is analyzed below. For one thing, from Assumption 7.6, there exists a corresponding $\boldsymbol{\alpha}_1^*$ such that

$$\|1_N^\top \boldsymbol{\alpha}_1^*(t) - (\mathbf{n}+\mathbf{1})_M^\top\| \text{ is minimized}, \forall t \geq T. \tag{7.22}$$

Furthermore, according to the condition of $M(n+1) \leq N$ in Assumption 7.3, it follows from Eq. (7.21) that there exists at least one target $h \neq l$ such that $|\mathbb{V}_h(t)| = \zeta_h(t) > n+1, \forall t \geq T$. Therefore, by reassigning at least one robot from the subgroup $\mathbb{V}_h$ to $\mathbb{V}_l$, we can always find another subgrouping matrix $\boldsymbol{\alpha}_2$ such that $\|1_N^\top \boldsymbol{\alpha}_2^* - (\mathbf{n}+\mathbf{1})_M^\top\| < \|1_N^\top \boldsymbol{\alpha}_1^* - (\mathbf{n}+\mathbf{1})_M^\top\|$, which contradicts Eq. (7.22). Hence, we conclude that $\mathbf{1}_N^\top \boldsymbol{\alpha} \geq (\mathbf{n}+\mathbf{1})_M^\top$ holds under Assumptions 7.3 and 7.6. □

Proposition 7.2 *Under Assumption 7.6, the MST fencing optimization problem in Eq. (7.19) ensures that robots do not experience infinite switching behaviors between different subgroups $\mathbb{V}_k$ for $k \in \mathcal{M}$. In other words, there cannot be multiple optimal solutions $\boldsymbol{\alpha}^*$ that repeatedly oscillate while achieving the minimum value of $\|\mathbf{1}_N^\top \boldsymbol{\alpha} - (\mathbf{n}+\mathbf{1})_M^\top\|$.*

Proof To prove this by contradiction, we start by assuming that there is a finite time $T_1 > T$ such that the optimal $\boldsymbol{\alpha}_1^*$ keeps invariant for $t \in (T, T_1)$ but not at $t = T_1$, which implies that there exist at least two optimal $\boldsymbol{\alpha}_1^* \neq \boldsymbol{\alpha}_2^*$ such that $\boldsymbol{\alpha}_1^* \to \boldsymbol{\alpha}_2^*$ at $t = T_1$ and satisfy the same minimum value, i.e.,

$$\|\mathbf{1}_N^{\mathrm{T}}\boldsymbol{\alpha}_1^*(T_1) - (\mathbf{n}+\mathbf{1})_M^{\mathrm{T}}\| = \|\mathbf{1}_N^{\mathrm{T}}\boldsymbol{\alpha}_2^*(T_1) - (\mathbf{n}+\mathbf{1})_M^{\mathrm{T}}\|. \tag{7.23}$$

For the elements $\alpha_{i,k}^{1*}$ in $\boldsymbol{\alpha}_1^*(t)$, when $t = T_1^-(i.e., t \to T_1^-)$, one has that the cost terms $\sum_{k=1}^{M}(\delta_{i,k}^r)^2$ in Eq. (7.19a) are minimized to be

$$\delta_{i,k}^r(T_1^-) > 0, \ (\text{since } \alpha_{i,k}^{1*}(T_1^-) = 1), \ \delta_{i,q}^r(T_1^-) = 0, \ \forall\, q \neq k, \tag{7.24}$$

while adhering to the constraints **C2** in Eq. (7.19b). At this time, $\delta_{i,k}^r(T_1^-) > 0$ signifies that robot i has approached the target k but has not yet fully converged to it.

However, if $\boldsymbol{\alpha}_1^* \to \boldsymbol{\alpha}_2^*$ at $t = T_1$, some robots are reallocated to different subgroups. For simplicity, assume that only robot i is reassigned to target l, such that $\{\alpha_{i,k}^{1*}(t) = 1, \alpha_{i,l}^{1*}(t) = 0\}$ transitions to $\{\alpha_{i,l}^{2*}(t) = 1, \alpha_{i,k}^{2*}(t) = 0\}$. Meanwhile, since robot i is closer to target k than to target l, it implies that robot i will require more energy u_i and a larger $\delta_{i,l}^r$ to meet the constraints **C2** in Eq. (7.19b) again. Thus,

$$\delta_{i,l}^r(T_1) > \delta_{i,k}^r(T_1^-), \ (\text{since } \alpha_{i,l}^{2*}(t) = 1), \ \delta_{i,q}^r(T_1) = 0, \ \forall\, q \neq l$$

with $\alpha_{i,l}^{2*}$ being a component of $\boldsymbol{\alpha}_2^*$. Although the values of $\|\mathbf{1}_N^{\mathrm{T}}\boldsymbol{\alpha}_2^*(T_1) - (\mathbf{n}+\mathbf{1})_M^{\mathrm{T}}\|$ and the collision-avoidance slack variables $\sum_{j\in\mathcal{N}_i}(\delta_{i,j}^c)^2$ remain constant, the cost terms $(\delta_{i,k}^r)^2 + \|u_i - v_k\|^2$ in Eq. (7.19) will increase at $t = T_1$. This contradicts the objective of minimizing the MST fencing optimization in Eq. (7.19). Therefore, the proof is concluded. □

7.5.2 Fencing Convergence

Given the MST fencing optimization in Eq. (7.19), once the optimal subgrouping matrix $\boldsymbol{\alpha}^*$ is established, the constraints **C1** and **C3** in Eq. (7.19b) for $\alpha_{i,q} = 0$ where $q \neq k \in \mathcal{M}$, as well as the constraints in Eq. (7.19d), can be omitted, which simplifies the cost function to $c\big((\delta_{i,k}^r)^2 + \sum_{j\in\mathcal{N}_i}(\delta_{i,j}^c)^2\big) + \|u_i - v_k\|^2$. Therefore, the optimization problem for robot i, $i \in \mathbb{V}_k)$ reduces to the following minimum-energy problem

$$\min_{u_i,\delta^r_{i,k},\delta^c_{i,j}} \left\{ c\Big((\delta^r_{i,k})^2 + \sum_{j\in\mathcal{N}_i}(\delta^c_{i,j})^2\Big) + \|u_i - v_k\|^2 \right\} \tag{7.25a}$$

$$\text{s.t. } \frac{\partial\phi^r_{i,k}(x_i)}{\partial x_i^{\mathsf{T}}}(u_i - v_k) + \gamma\big(\phi^r_{i,k}(x_i)\big) - \delta^r_{i,k} \le 0, \tag{7.25b}$$

$$\frac{\partial\phi^c_{i,j}(x_i)}{\partial x_i^{\mathsf{T}}}(u_i - v_k) + \gamma\big(\phi^c_{i,j}(x_i)\big) - \delta^c_{i,j} \le 0,\ j \in \mathcal{N}_i(t). \tag{7.25c}$$

Lemma 7.1 *The robots within subgroup $\mathbb{V}_k$, as managed by the minimum-energy optimization problem in Eq. (7.25), achieve the Objective 3 of Definition 7.3, which guarantees that collisions between robots are consistently avoided, i.e., $\|x_i(t) - x_j(t)\| > r, \forall t > 0, i \ne j \in \mathbb{V}_k$.*

Proof Let N_i represent the number of elements in the set $\mathcal{N}_i$, i.e., $N_i = |\mathcal{N}_i|$. We define

$$\begin{aligned}
\boldsymbol{\delta}_i &= [\delta^r_{i,k}, \overbrace{\ldots, \delta^c_{i,j}, \ldots}^{j\in\mathcal{N}_i}]^{\mathsf{T}} \in \mathbb{R}^{N_i+1},\\
\Phi_i &= [\phi^r_{i,k}(x_i), \ldots, \phi^c_{i,j}(x_i), \ldots]^{\mathsf{T}} \in \mathbb{R}^{N_i+1},\\
\gamma(\Phi_i) &= \big[\gamma\big(\phi^r_{i,k}(x_i)\big), \ldots, \gamma\big(\phi^c_{i,j}(x_i)\big), \ldots\big]^{\mathsf{T}} \in \mathbb{R}^{N_i+1}.
\end{aligned} \tag{7.26}$$

Combining with Eqs. (7.25) and (7.26) yields

$$\min_{u_i,\boldsymbol{\delta}_i} \left\{ c\boldsymbol{\delta}_i^{\mathsf{T}}\boldsymbol{\delta}_i + \|u_i - v_k\|^2 \right\},\ \text{s.t. } \mathbf{g}_i \le \mathbf{0}_{N_i+1}, \tag{7.27}$$

with $\mathbf{g}_i := [g_{i,1}, \ldots, g_{i,N_i+1}]^{\mathsf{T}} = \frac{\partial\Phi_i(x_i)}{\partial x_i^{\mathsf{T}}}(u_i - v_k) + \gamma(\Phi_i(x_i)) - \boldsymbol{\delta}_i \in \mathbb{R}^{N_i+1}$. Let $\boldsymbol{\lambda}_i := [\lambda_{i,1}, \ldots, \lambda_{i,N_i+1}]$ where each entry is nonnegative, one has that the corresponding Lagrangian function for Eq. (7.27) is $\mathcal{L}_i(u_i, \boldsymbol{\delta}_i, \boldsymbol{\lambda}_i) = c\boldsymbol{\delta}_i^{\mathsf{T}}\boldsymbol{\delta}_i + \|u_i - v_k\|^2 + \boldsymbol{\lambda}_i^{\mathsf{T}}\mathbf{g}_i$. Using Remark 7.2, it follows from the KKT conditions [26] and Eq. (7.27) that

$$\frac{\partial\mathcal{L}_i(u_i^*, \boldsymbol{\delta}_i^*, \boldsymbol{\lambda}_i^*)}{\partial[u_i^{\mathsf{T}}, \boldsymbol{\delta}_i^{\mathsf{T}}]^{\mathsf{T}}} = \mathbf{0}_{n+N_i+1}, (\boldsymbol{\lambda}_i^*)^{\mathsf{T}}\mathbf{g}_i = 0, \boldsymbol{\lambda}_i^* \ge \mathbf{0}_{N_i+1}, \mathbf{g}_i \le \mathbf{0}_{N_i+1}, \tag{7.28}$$

where $u_i^*, \boldsymbol{\delta}_i^*, \boldsymbol{\lambda}_i^*$ are the optimal parameters. Since the Lagrangian multipliers $\lambda_{i,1}$ and $\lambda_{i,l}, l \in \mathbb{Z}_2^{N_i+1}$ are designed for the different constraints (7.25b)–(7.25c), respectively, we will analyze the collision avoidance in the following four cases.

Case 1: $\boldsymbol{\lambda}_i^* = \mathbf{0}_{N_i+1}, i \in \mathcal{V}$.

If Case 1 holds, it follows from $(\boldsymbol{\lambda}_i^*)^{\mathsf{T}}\mathbf{g}_i = 0$ in Eq. (7.28) that the boundary of the constraints $\mathbf{g}_i \le \mathbf{0}_{N_i+1}$ are not necessarily activated. Additionally, one has that $2(u_i^* - v_k) = \mathbf{0}_n$, $2c\boldsymbol{\delta}_i^* = \mathbf{0}_{N_i+1}$, which implies that $u_i^* = v_k$, $\boldsymbol{\delta}_i^* = \mathbf{0}_{N_i+1}$. Recalling (7.26) and substituting $u_i^*, \boldsymbol{\delta}_i^*$ into the constraints (7.25b)–(7.25c) yields $\gamma\big(\phi^r_{i,k}(x_i)\big) \le$

$0 \Rightarrow \phi^r_{i,k}(x_i) \leq 0$, and $\gamma\left(\phi^c_{i,j}(x_i)\right) \leq 0,\ j \in \mathcal{N}_i, \Rightarrow \phi^c_{i,j}(x_i) \leq 0$, which then follows from Eqs. (7.11), (7.16) and (7.17) that $x_i = r_k,\ i \in \mathcal{V}_k,\ \|x_i - x_j\| \geq \rho > r,\ j \in \mathcal{N}_i$. Then, the neighboring inter-robot collision avoidance is naturally guaranteed in Case 1.

Case 2: $\lambda^*_{i,1} > 0, \lambda^*_{i,l} = 0, \forall l \in \mathbb{Z}_2^{N_i+1} (\boldsymbol{\lambda}^*_i \neq \mathbf{0}_{N_i+1})$.

If Case 2 holds, it follows from the relationships of $\lambda_{i,l}$ and x_i in (7.6) that $\lambda^*_{i,l} = 0, \forall l \in \mathbb{Z}_2^{N_i+1} \Rightarrow \phi^c_{i,j}(x_i) \leq 0, \forall j \in \mathcal{N}_i$, which implies that $\|x_i - x_j\| \geq \rho > r,\ j \in \mathcal{N}_i$, i.e., the neighboring collision avoidance is guaranteed in Case 2.

Case 3: $\lambda^*_{i,1} = 0, \lambda^*_{i,j} > 0, \exists j \in \mathbb{Z}_2^{N_i+1} (\boldsymbol{\lambda}^*_i \neq \mathbf{0}_{N_i+1})$.

For $\lambda^*_{i,q} = 0, q \neq j \in \mathbb{Z}_2^{N_i+1}$, analogous to Case 2, it follows from Eq. (7.6) that $\|x_i - x_q\| \geq \rho > r, q \in \mathcal{N}_i$, then collision avoidance between robots i and q is satisfied.

For $\lambda^*_{i,j} > 0$, let $\mathcal{A}_i \neq \emptyset \subseteq \mathcal{N}_i$ for which $\lambda^*_{i,j} > 0$, and a_i be the cardinality of the set $\mathcal{A}_i$. We define $\widetilde{\boldsymbol{\lambda}}_i := [\lambda_{i,1}, \ldots, \lambda_{i,l}, \ldots]^\top \in \mathbb{R}^{a_i+1}$, $\widetilde{\mathbf{g}}_i := [g_{i,1}, \ldots, g_{i,l}, \ldots]^\top \in \mathbb{R}^{a_i+1}$, $\boldsymbol{\delta}_i = [\delta_{i,k}, \ldots, \delta^c_{i,j}]^\top \in \mathbb{R}^{N_i+1}$, $\widetilde{\Phi}_i(x_i) = [\phi^r_{i,k}, \ldots, \phi^c_{i,j}, \ldots]^\top \in \mathbb{R}^{a_i+1}$, $\gamma(\widetilde{\Phi}_i(x_i)) = \left[\gamma\left(\phi^r_{i,k}(x_i)\right), \ldots, \gamma\left(\phi^c_{i,j}(x_i)\right), \ldots\right]^\top \in \mathbb{R}^{a_i+1}$ one has that the Lagrangian function for Eq. (7.27) becomes $\mathcal{L}_i(u_i, \widetilde{\boldsymbol{\delta}}_i^{[3]}, \widetilde{\boldsymbol{\lambda}}_i^{[3]}) = c(\widetilde{\boldsymbol{\delta}}_i^{[3]})^\top \widetilde{\boldsymbol{\delta}}_i^{[3]} + \|u_i - v_k\|^2 + (\widetilde{\boldsymbol{\lambda}}_i^{[3]})^\top \widetilde{\mathbf{g}}_i^{[3]}$, which implies that the stationarity condition in Eq. (7.28) becomes

$$\begin{bmatrix} 2(u^*_i - v_k) \\ 2c\widetilde{\boldsymbol{\delta}}_i^{[3]*} \end{bmatrix} + \begin{bmatrix} \frac{\partial (\widetilde{\Phi}_i^{[3]})^\top}{\partial x_i} \widetilde{\boldsymbol{\lambda}}_i^{[3]*} \\ -\widetilde{\boldsymbol{\lambda}}_i^{[3]*} \end{bmatrix} = \mathbf{0}_{n+N_i+1}, \tag{7.29}$$

which implies that

$$u^*_i = v_k - \frac{\partial (\widetilde{\Phi}_i^{[3]})^\top}{\partial x_i} \frac{\widetilde{\boldsymbol{\lambda}}_i^{[3]*}}{2}, \quad \widetilde{\boldsymbol{\delta}}_i^{[3]*} = \frac{1}{2c} \widetilde{\boldsymbol{\lambda}}_i^{[3]*}. \tag{7.30}$$

Substituting u^*_i in Eq. (7.30) into $\widetilde{\mathbf{g}}_i = \mathbf{0}_{a_i+1}$ in Eqs. (7.27), (7.29) yields $\widetilde{\boldsymbol{\delta}}_i^{[3]*} = \Xi_i^{[3]} \gamma(\widetilde{\Phi}_i^{[3]})$ with

$$\Xi_i^{[3]} = \left(I_{a_i} + c \frac{\partial \widetilde{\Phi}_i^{[3]}}{\partial x_i^\top} \frac{\partial (\widetilde{\Phi}_i^{[3]})^\top}{\partial x_i} \right)^{-1} \in \mathbb{R}^{a_i \times a_i}. \tag{7.31}$$

It, together with Eq. (7.30), gives that u^*_i

$$u^*_i = v_k - c \frac{\partial (\widetilde{\Phi}_i^{[3]})^\top}{\partial x_i} \Xi_i^{[3]} \gamma(\widetilde{\Phi}_i^{[3]}). \tag{7.32}$$

Based on u^*_i in Eq. (7.32) and $\gamma(\widetilde{\Phi}_i^{[3]})$ in Eq. (7.29), we pick a Lyapunov function candidate

$$V_1(t) = \frac{1}{2}\sum_{i\in\mathbb{V}_k}\sum_{j\in\mathcal{A}_i}\gamma(\phi_{i,j}^c(x_i))^2. \tag{7.33}$$

By choosing $\gamma(\phi) = \phi$ as a simple choice, the derivative of $V(t)$, as described in Eq. (7.33), can be derived from Eqs. (7.11), (7.16), and (7.33), as follows:

$$\dot{V}_1(t) = \sum_{i\in\mathbb{V}_k}\sum_{j\in\mathcal{A}_i}\phi_{i,j}^c(x_i)\frac{\partial\phi_{i,j}^c(x_i)}{\partial x_i^\top}(u_i - u_j) \tag{7.34}$$

with $\partial\phi_{i,j}^c(x_i)/\partial x_i^\top = \partial\phi_{i,j}^c(x_i)/\partial(x_i - x_j)^\top$. Using the fact that

$$\sum_{i\in\mathbb{V}_k}\sum_{j\in\mathcal{A}_i}\phi_{i,j}^c(x_i)\frac{\partial\phi_{i,j}^c(x_i)}{\partial x_i^\top}(u_i - u_j) = 2\sum_{i\in\mathbb{V}_k}\sum_{j\in\mathcal{A}_i}\phi_{i,j}^c(x_i)\frac{\partial\phi_{i,j}^c(x_i)}{\partial x_i^\top}(u_i - v_k), \tag{7.35}$$

it, together with Eqs. (7.29), (7.34), gives that

$$\dot{V}_1(t) = 2\sum_{i\in\mathbb{V}_k}(\widetilde{\Phi}_i^{[3]})^\top\frac{\partial\widetilde{\Phi}_i^{[3]}}{\partial x_i^\top}(u_i - v_k). \tag{7.36}$$

Substituting u_i^* in Eq. (7.32) into Eq. (7.36) yields

$$\dot{V}_1(t) = -2c\sum_{i\in\mathbb{V}_k}(\widetilde{\Phi}_i^{[3]})^\top\frac{\partial\widetilde{\Phi}_i^{[3]}}{\partial x_i^\top}\frac{\partial(\widetilde{\Phi}_i^{[3]})^\top}{\partial x_i}\Xi_i^{[3]}\widetilde{\Phi}_i^{[3]} \tag{7.37}$$

with $\gamma(\widetilde{\Phi}_i^{[3]}) = \widetilde{\Phi}_i^{[3]}$. Additionally, based on Woodbury matrix identity [27], $\Xi_i^{[3]}$ in Eq. (7.31) is formulated to be

$$\Xi_i^{[3]} = I_{a_i} - c\frac{\partial\widetilde{\Phi}_i^{[3]}}{\partial x_i^\top}\widetilde{\Xi}_i^{[3]}\frac{\partial(\widetilde{\Phi}_i^{[3]})^\top}{\partial x_i} \tag{7.38}$$

with

$$\widetilde{\Xi}_i^{[3]} = \left(I_n + c\frac{\partial(\widetilde{\Phi}_i^{[3]})^\top}{\partial x_i}\frac{\partial\widetilde{\Phi}_i^{[3]}}{\partial x_i^\top}\right)^{-1} \in \mathbb{R}^{n\times n}, \tag{7.39}$$

which follows from Eqs. (7.38) and (7.39) that

$$\frac{\partial(\widetilde{\Phi}_i^{[3]})^\top}{\partial x_i}\Xi_i^{[3]} = \widetilde{\Xi}_i^{[3]}\frac{\partial(\widetilde{\Phi}_i^{[3]})^\top}{\partial x_i}. \tag{7.40}$$

Substituting Eq. (7.40) into Eq. (7.37) yields

$$\dot{V}_1(t) = -2c \sum_{i\in\mathbb{V}_k} (\widetilde{\Phi}_i^{[3]})^{\mathsf{T}} \frac{\partial \widetilde{\Phi}_i^{[3]}}{\partial x_i^{\mathsf{T}}} \widetilde{\Xi}_i^{[3]} \frac{\partial (\widetilde{\Phi}_i^{[3]})^{\mathsf{T}}}{\partial x_i} \widetilde{\Phi}_i^{[3]}. \tag{7.41}$$

Given that $\widetilde{\Xi}_i^{[3]}$ in Eq. (7.39) is positive definite, it follows that its smallest eigenvalue is strictly positive, i.e., $\lambda_{\min}(\widetilde{\Xi}_i^{[3]}) > 0$, which indicates that

$$\dot{V}_1(t) \leq -2c\lambda_{\min}(\widetilde{\Xi}_i^{[3]}) \sum_{i\in\mathbb{V}_k} \left\| (\widetilde{\Phi}_i^{[3]})^{\mathsf{T}} \frac{\partial \widetilde{\Phi}_i^{[3]}}{\partial x_i^{\mathsf{T}}} \right\|^2 \leq 0. \tag{7.42}$$

From Assumption 7.4 and the definition of V_1 in Eq. (7.33), it is evident that the initial value of $V_1(0)$ is bounded. Combined with the fact that $\dot{V}_1 \leq 0$ as given in Eq. (7.42), this implies that $V(t)$ is bounded for all $t > 0$ in Case 3. Consequently, $\phi_{i,j}^c(x_i)$ is also bounded. Furthermore, from Eqs. (7.17) and (7.33), it follows that $\phi_{i,j}^c(x_i) \to \infty$, if $\|x_i - x_j\| \to r$, $j \in \mathcal{A}_i$. This indicates that $\|x_i - x_j\| > r, \forall j \in \mathcal{A}_i$ at all times, which thereby ensures neighboring inter-robot collision avoidance in Case 3.

Case 4: $\lambda_{i,1}^* > 0, \lambda_{i,j}^* > 0, \exists j \in \mathbb{Z}_2^{N_i+1} (\boldsymbol{\lambda}_i^* \neq \mathbf{0}_{N_i+1})$.

The approach to neighboring collision avoidance in Case 4 mirrors that of Case 3. In particular, when $\lambda_{i,q}^* = 0$ for $q \neq j \in \mathbb{Z}_2^{N_i+1}$, and if $\|x_i - x_q\| \geq \rho > r$ for $q \in \mathcal{N}_i$, then the collision avoidance between robots i and q is assured.

For $\lambda_{i,1}^* > 0, \lambda_{i,j}^* > 0$, we define $\widetilde{\boldsymbol{\lambda}}_i^{[4]} := [\lambda_{i,1}, \ldots, \lambda_{i,l}, \ldots]^{\mathsf{T}} \in \mathbb{R}^{a_i+1}, l \in \mathcal{A}_i$, $\widetilde{\mathbf{g}}_i^{[4]} := [g_{i,1}, \ldots, \quad g_{i,l}, \ldots]^{\mathsf{T}} \in \mathbb{R}^{a_i+1}$, $\widetilde{\boldsymbol{\delta}}_i^{[4]} = [\delta_{i,k}^r, \ldots, \delta_{i,j}^c, \ldots]^{\mathsf{T}} \in \mathbb{R}^{a_i+1}$, $\widetilde{\Phi}_i^{[4]} := [\phi_{i,k}^r(x_i), \ldots, \phi_{i,j}^c(x_i), \ldots]^{\mathsf{T}} \in \mathbb{R}^{a_i+1}$, $\gamma(\widetilde{\Phi}_i^{[4]}) := \big[\gamma\big(\phi_{i,k}^r(x_i)\big), \ldots, \gamma\big(\phi_{i,j}^c(x_i)\big), \ldots\big]^{\mathsf{T}} \in \mathbb{R}^{a_i+1}$ with $\mathcal{A}_i$ and a_i given in Case 3, which then formulates the Lagrangian function in Eq. (7.27) to be

$$\mathcal{L}_i(u_i, \widetilde{\boldsymbol{\delta}}_i^{[4]}, \widetilde{\boldsymbol{\lambda}}_i^{[4]}) = c(\widetilde{\boldsymbol{\delta}}_i^{[4]})^{\mathsf{T}} \widetilde{\boldsymbol{\delta}}_i^{[4]} + \|u_i - v_k\|^2 + (\widetilde{\boldsymbol{\lambda}}_i^{[4]})^{\mathsf{T}} \widetilde{\mathbf{g}}_i^{[4]}.$$

Analogous to Eq. (7.29), one has that the optimal u_i^* is calculated to be

$$u_i^* = v_k - c\frac{\partial (\widetilde{\Phi}_i^{[4]})^{\mathsf{T}}}{\partial x_i} \Xi_i^{[4]} \gamma(\widetilde{\Phi}_i^{[4]}) \tag{7.43}$$

with

$$\Xi_i^{[4]} = \left(I_{a_i+1} + c\frac{\partial \widetilde{\Phi}_i^{[4]}}{\partial x_i^{\mathsf{T}}} \frac{\partial (\widetilde{\Phi}_i^{[4]})^{\mathsf{T}}}{\partial x_i} \right)^{-1} \in \mathbb{R}^{(a_i+1)\times(a_i+1)}. \tag{7.44}$$

Then, we pick a Lyapunov function candidate

$$V_2(t) = \sum_{i \in \mathbb{V}_k} \gamma(\phi^r_{i,k}(x_i))^2 + \frac{1}{2} \sum_{i \in \mathbb{V}_k} \sum_{j \in \mathcal{A}_i} \gamma(\phi^c_{i,j}(x_i))^2. \tag{7.45}$$

Similarly to the proof in Case 3, from Eqs. (7.34) to (7.42), the derivative of $V_2(t)$ in Eq. (7.45) is given by

$$\dot{V}_2(t) \leq -2c\lambda_{\min}(\widetilde{\Xi}_i^{[4]}) \sum_{i \in \mathbb{V}_k} \left\| (\widetilde{\Phi}_i^{[4]})^{\mathsf{T}} \frac{\partial \widetilde{\Phi}_i^{[4]}}{\partial x_i^{\mathsf{T}}} \right\|^2 \tag{7.46}$$

with the positive-definite matrix $\widetilde{\Xi}_i^{[4]}$ being

$$\widetilde{\Xi}_i^{[4]} = \left(I_n + c \frac{\partial (\widetilde{\Phi}_i^{[4]})^{\mathsf{T}}}{\partial x_i} \frac{\partial \widetilde{\Phi}_i^{[4]}}{\partial x_i^{\mathsf{T}}} \right)^{-1} \in \mathbb{R}^{n \times n}. \tag{7.47}$$

Since $\dot{V}_2 \leq 0$ as shown in Eq. (7.46), the analysis for neighboring collision avoidance in Case 4, which requires $\|x_i - x_j\| > r$ for all $j \in \mathcal{A}_i$, is analogous to that in Case 3 and is therefore not repeated. Consequently, by integrating the results from Cases 1 through 4, the proof for collision avoidance is completed. □

Lemma 7.2 *The robots in subgroup $\mathbb{V}_k$, optimized through the minimum-energy approach in Eq. (7.25), will ultimately achieve Objective 1 as defined in Definition 7.3. Specifically, each target r_k will be fenced within the convex hull $\mathrm{co}(\mathbb{V}_k)$, i.e., $\lim_{t\to\infty} \mathrm{dist}\big(r_k(t), \mathrm{co}(\mathbb{V}_k(t))\big) = 0, \forall k \in \mathcal{M}$.*

Proof Similar to the proof of Lemma 7.1, we will analyze the convergence of the flexible-ordering convex hull by considering four distinct cases.

Case 1: $\boldsymbol{\lambda}_i^* = \mathbf{0}_{N_i+1}, i \in \mathcal{V}$.

In Case 1, referring to Case 1 of Lemma 7.1, we have $x_i = r_k$ for $i \in \mathcal{V}_k$, $\|x_i - x_j\| \geq \rho > r$, $j \in \mathcal{N}_i$. This indicates that robot i approaches target k and remains there. However, this scenario cannot be sustained continuously for all robots in $\mathbb{V}_k$, as it would inevitably conflict with the collision avoidance requirements of the other robots. Therefore, Case 1 is excluded.

Case 2: $\lambda_{i,1}^* > 0, \lambda_{i,l}^* = 0, \forall l \in \mathbb{Z}_2^{N_i+1} (\boldsymbol{\lambda}_i^* \neq \mathbf{0}_{N_i+1})$.

In Case 2, it follows from $(\boldsymbol{\lambda}_i^*)^{\mathsf{T}} \mathbf{g}_i = 0$ in Eq. (7.28) that only the boundary of the constraint $g_{i,1} \leq 0$ is activated, i.e., $g_{i,1} = 0$. Then, from the stationarity condition in (7.28) and $\boldsymbol{\lambda}_i$, $\boldsymbol{\delta}_i$ in (7.26), one has that

$$2(u_i^* - v_k) = -\frac{\partial \phi^r_{i,k}(x_i)}{\partial x_i} \lambda_{i,1}^*, \; 2c\delta_{i,k}^{r*} = \lambda_{i,1}^*, \; 2c\delta_{i,j}^{c*} = 0, \forall j \in \mathcal{N}_i. \tag{7.48}$$

Substituting u_i^*, $\delta_{i,k}^{r*}$ in (7.48) into $g_{i,1} = 0$ yields

$$-c \frac{\partial \phi^r_{i,k}(x_i)}{\partial x_i^{\mathsf{T}}} \frac{\partial \phi^r_{i,k}(x_i)}{\partial x_i} \delta_{i,k}^{r*} + \gamma\big(\phi^r_{i,k}(x_i)\big) - \delta_{i,k}^{r*} = 0,$$

one has that $\delta_{i,k}^{r*} = \gamma\big(\phi_{i,k}^{r}(x_i)\big)\varsigma_i^{-1}$ with

$$\varsigma_i := c\frac{\partial \phi_{i,k}^{r}(x_i)}{\partial x_i^{\mathsf{T}}}\frac{\partial \phi_{i,k}^{r}(x_i)}{\partial x_i} + 1 > 0.$$

Therefore, it can be deduced that the optimal input u_i^* is formulated to be

$$u_i^* = v_k - c\frac{\partial \phi_{i,k}^{r}(x_i)}{\partial x_i}\gamma\big(\phi_{i,k}^{r}(x_i)\big)\varsigma_i^{-1},$$

which implies that robot i, for $i \in \mathbb{V}_k$, converges to the target k such that $x_i = r_k$. Consequently, Case 2 would essentially be reduced to Case 1, which has already been excluded. Thus, Case 2 is also excluded.

Case 3: $\lambda_{i,1}^* = 0, \lambda_{i,j}^* > 0, \exists j \in \mathbb{Z}_2^{N_i+1}(\boldsymbol{\lambda}_i^* \neq \mathbf{0}_{N_i+1})$.

In Case 3, given that $\lambda_{i,1}^* = 0$ and from Eq. (7.6), we have $x_i \in \mathcal{T}_{i,k}^{r}$, indicating that robot i is positioned exactly at target k. However, this condition cannot be sustained for the entire robot subgroup $\mathbb{V}_k$. At least one other robot, $q \neq i \in \mathcal{V}$, would also need to be at the position of target k, which directly conflicts with the collision avoidance requirements between robots i and q. Thus, Case 3 is also not feasible, analogous to Case 1.

Case 4: $\lambda_{i,1}^* > 0, \lambda_{i,j}^* > 0, \exists j \in \mathbb{Z}_2^{N_i+1}(\boldsymbol{\lambda}_i^* \neq \mathbf{0}_{N_i+1})$.

In Case 4, recalling $\dot{V}_2(t)$ in (7.46), integrating both sides of Eq. (7.46) in the time interval $t \in [T, \infty)$ yields $V_2(t) \leq V_2(T) - \int_{t_2}^{\infty}\sigma(s)ds$ with

$$\sigma := 2c\lambda_{\min}(\widetilde{\Xi}_i^{[4]})\sum_{i\in\mathbb{V}_k}\left\|(\widetilde{\Phi}_i^{[4]})^{\mathsf{T}}\frac{\partial\widetilde{\Phi}_i^{[4]}}{\partial x_i^{\mathsf{T}}}\right\|^2. \tag{7.49}$$

Recalling Lemma 7.1, it follows from (7.46) that $V_2(T)$ is bounded. Meanwhile, since $-\int_{t_2}^{\infty}\sigma(s)ds \leq 0$, one has that the states $\gamma(\phi_{i,k}^{r}(x_i)), \gamma(\phi_{i,j}^{c}(x_i)), \phi_{i,k}^{r}(x_i), \phi_{i,j}^{c}(x_i)$ are bounded as well, which further implies that $\int_{t_2}^{\infty}\sigma(s)ds$ is upper bounded. Then, it follows from Eqs. (7.11), (7.16), (7.49) that $\ddot{\sigma}$ are bounded as well, which implies that $\dot{\sigma}$ is uniformly continuous in $t \in [T, \infty)$. According to Barbalat's Lemma [28], one has that $\lim_{t\to\infty}\sigma(t) = 0$, i.e.,

$$\lim_{t\to\infty}\left\|(\widetilde{\Phi}_i^{[4]}(t))^{\mathsf{T}}\frac{\partial\widetilde{\Phi}_i^{[4]}(t))}{\partial x_i^{\mathsf{T}}(t)}\right\| = 0. \tag{7.50}$$

With $\widetilde{\Phi}_i^{[4]}$ given in Eq. (7.43), it follows from Eqs. (7.11) and (7.16) that

$$\lim_{t\to\infty}\left\{x_i(t) - r_k(t) - \sum_{j\in\mathcal{A}_i}\left(\frac{1}{\|x_i(t)-x_j(t)\|-r} - \frac{1}{\rho-r}\|\right)\right.$$
$$\left.\frac{x_i(t)-x_j(t)}{\|x_i(t)-x_j(t)\|^3}\right\} = \mathbf{0}_n, \forall i \in \mathbb{V}_k. \quad (7.51)$$

From the fact that

$$\sum_{i\in\mathbb{V}_k}\sum_{j\in\mathcal{A}_i}\left(\frac{1}{\|x_i(t)-x_j(t)\|-r} - \frac{1}{(\rho-r)}\right)\frac{x_i(t)-x_j(t)}{\|x_i(t)-x_j(t)\|^3} = \mathbf{0}_n,$$

summing (7.51) in the subgroup of $\mathbb{V}_k$ yields

$$\lim_{t\to\infty}\sum_{i\in\mathbb{V}_k}\{x_i(t) - r_k(t)\} = \lim_{t\to\infty}\sum_{i\in\mathbb{V}_k}\frac{(x_i(t)-r_k(t))}{N} = \mathbf{0}_n.$$

Let $\mathrm{co}(\mathbb{V}_k) = 1/N\sum_{i\in\mathbb{V}_k} x_i(t)$, it follows from $\sum_{i\in\mathbb{V}_k} r_k/N = r_k$ that

$$\lim_{t\to\infty}\mathrm{dist}\big(r_k(t), \mathrm{co}(\mathbb{V}_k(t))\big) = \lim_{t\to\infty}\left\{\frac{1}{N}\sum_{i\in\mathbb{V}_k} x_i(t) - r_k(t)\right\} = \mathbf{0}_n,$$

which implies that r_k will be enclosed within the convex hull $\mathrm{co}(\mathbb{V}_k) = 1/N\sum_{i\in\mathbb{V}_k} x_i(t)$, fulfilling Objective 1 as defined in Definition 7.3. Therefore, by considering Cases 1 through 4, the proof is completed. □

Lemma 7.3 *The robots in the subgroup $\mathbb{V}_k$, managed by the minimum-energy optimization* (7.25)*, will eventually accomplish Objective 2 as defined in Definition 7.3. In particular, the relative velocities between any robot i (where $i \in \mathbb{V}_k$) and the target v_k will asymptotically diminish to zero, i.e.,* $\lim_{t\to\infty}\{u_i(t) - v_k(t)\} = \mathbf{0}_n, i \in \mathbb{V}_k$.

Proof Analogous to the proof of Lemma 7.2, we can exclude Cases 1–3 and focus on the convergence analysis in Case 4. On one hand, analogous to Eq. (7.40) in Lemma 7.2, one has that $(\partial(\widetilde{\Phi}_i^{[4]})^{\mathsf{T}}/\partial x_i)\Xi_i^{[4]} = \widetilde{\Xi}_i^{[4]}(\partial(\widetilde{\Phi}_i^{[4]})^{\mathsf{T}}/\partial x_i)$ with $\Xi_i^{[4]}$, $\widetilde{\Xi}_i^{[4]}$, $\widetilde{\Phi}_i^{[4]}$ given in (7.43), (7.44), (7.47), which implies that u_i^* in (7.43) can be reformulated to be

$$u_i^* = v_k - c\widetilde{\Xi}_i^{[4]}\frac{\partial(\widetilde{\Phi}_i^{[4]})^{\mathsf{T}}}{\partial x_i}\gamma(\widetilde{\Phi}_i^{[4]}). \quad (7.52)$$

On the other hand, it follows from Eqs. (7.47) and (7.50) that

$$\lim_{t\to\infty}\widetilde{\Xi}_i^{[4]}(t)\frac{\partial(\widetilde{\Phi}_i^{[4]}(x_i(t)))^{\mathsf{T}}}{\partial x_i(t)}\gamma(\widetilde{\Phi}_i^{[4]}(x_i(t))) = \mathbf{0}_n. \quad (7.53)$$

Combining Eqs. (7.52) and (7.53) together, one has that $\lim_{t\to\infty} u_i^*(t) = v_k$, which implies that Objective (2) in Definition 7.3 is achieved. The proof is thus completed. □

Theorem 7.1 *Under Assumptions 7.1–7.6, all robots governed by* (7.2) *and* (7.19) *successfully achieve the flexible subgrouping and flexible-ordering fencing of the moving targets governed by* (7.7). *In other words, Conditions (1)–(2) in Definition 7.2 and Objectives (1)–(3) in Definition 7.3 are satisfied.*

Proof We draw the conclusion by Propositions 7.1–7.2, and Lemmas 7.1–7.3 directly. □

Remark 7.3 Although the FSOC algorithm (7.19) may not be directly suitable for vehicles with general affine dynamics, it can be extended to various types of vehicles through a hierarchical control approach, such as unicycles, unmanned aerial vehicles (UAVs), and unmanned surface vessels (USVs) [23]. In this setup, the optimal commands u_i^* from (7.19) serve as higher-level guidance velocities. As long as the low-level velocity tracking system can effectively follow the desired velocities in real-time, the FSOC algorithm remains applicable to many real-world robotic systems. Moreover, by defining obstacle sets and incorporating obstacle-avoidance constraints similar to those in Eqs. (7.15) and (7.18), the proposed FSOC algorithm (7.19) can be adapted for MST fencing in environments with obstacles. This will be demonstrated through subsequent 2D and 3D simulations.

Remark 7.4 The FSOC algorithm (7.19) has a few notable limitations. (i) When targets are positioned very close to one another, the robots' collision-avoidance behavior may disrupt the convex-hull fencing process. (ii) If the robots' dynamics are modeled as general affine systems, the algorithm could become ineffective due to the complexity in choosing suitable CBFs. (iii) The algorithm might not perform well in environments with polygonal obstacles instead of circular ones, as the current CBF constraints are designed for smooth obstacle boundaries. (iv) Unpredictable target dynamics could hinder the robots' ability to fence the targets effectively. As a preliminary approach to the MST fencing problem, further research is needed to overcome these challenges and improve the algorithm's robustness.

7.6 Simulation Results

This section presents a series of simulations to verify the flexibility, robustness, adaptability, and scalability of the FSOC algorithm. Initially, we employ single integrator dynamics in Eq. (7.2) within the control barrier function framework to guarantee the asymptotic convergence of the FSOC algorithm (7.19). Although these dynamics may not directly suit robots with more complex models, the FSOC algorithm (7.19) can still be applied to a practical unicycle model for the robot group $\mathcal{V}$, as demonstrated below [23].

$$\dot{x}_{i,1} = v_{i,l} \cos\theta_i, \dot{x}_{i,2} = v_{i,l} \sin\theta_i, \dot{\theta}_i = v_{i,a},$$
$$(\dot{x}_{i,3} = v_{i,c}, \text{ if robots are in 3D}), i \in \mathcal{V}, \tag{7.54}$$

where $[x_{i,1}, x_{i,2}, x_{i,3}]^\top \in \mathbb{R}^3$ and $\theta_i \in \mathbb{R}$ represent the position and X-Y heading angle of robot i, respectively, while $v_{i,l} \in \mathbb{R}$, $v_{i,a} \in \mathbb{R}$, and $v_{i,c} \in \mathbb{R}$ denote its linear, angular, and climbing velocities. Utilizing the near-identity diffeomorphism [29], the optimal FSOC control inputs u_i^* derived in (7.19) can be efficiently mapped to the desired velocities $v_{i,l}, v_{i,a}, v_{i,c}$ for the practical unicycle model presented in (7.54) below

$$v_{i,l} = u_{i,1}^* \cos\theta_i + u_{i,2}^* \sin\theta_i, v_{i,a} = -\frac{u_{i,1}^*}{l} \sin\theta_i + \frac{u_{i,2}^*}{l} \cos\theta_i,$$
$$(v_{i,c} = u_{i,3}^* \text{ if robots are in 3D}), i \in \mathcal{V},$$

where $u_i^* = [u_{i,1}^*, u_{i,2}^*, u_{i,3}^*]^\top$, and $l \in \mathbb{R}^+$ represents a small positive constant. The sensing, desired, and collision radii are chosen as $R = 6$ m, $\rho = 4$ m, and $r = 1.5$ m, respectively. Under Assumption 7.5, the weights in the FSOC optimization (7.19) are set to $b = 100{,}000$ and $c = 100$.

To demonstrate the flexibility of the proposed FSOC algorithm (7.19), we run a simulation involving ten robots starting from different initial positions and two scattered targets. As shown in Fig. 7.4a, b, the robots successfully form pentagon formations around the targets. Although the overall shapes are similar, the specific subgroups and ordering sequences differ, highlighting the flexibility and high efficiency of the FSOC algorithm (7.19).

In addition, Fig. 7.4a provides a detailed view of the state evolution. As illustrated in Fig. 7.5, the flexible-ordering fencing errors, $1/N \sum_{i \in \mathbb{V}_k} x_i - r_k$ for $k = 1, 2$, converge to zero, which means that $\lim_{t\to\infty} \text{dist}\big(r_k(t), \text{co}(\mathbb{V}_k(t))\big) = \mathbf{0}_n$, thereby ensuring the achievement of Objective (1) in Definition 7.3. Figure 7.6 shows that the velocity-tracking errors, $u_i - v_k$ for all $i \in \mathbb{V}_k$, also converge to zero, confirming that Objective (2) in Definition 7.3 is met. Furthermore, Fig. 7.7 reveals that the inter-robot distances, $\|x_i - x_j\| > 1.5$ for $i \neq j \in \mathcal{V}$, which ensures that the collision avoidance requirement in Objective (3) of Definition 7.3 is satisfied. For comparison, we also simulated the fixed subgroup + fencing algorithm [19], as shown in Fig. 7.4c, d. These results show that robots under the fixed subgroup method travel longer distances compared to those using the FSOC algorithm in Fig. 7.4a, b.

To demonstrate the robustness of the proposed FSOC algorithm (7.19), we analyze two scenarios: one where some robots suddenly break down and another where a new target appears during the fencing process. In the first scenario, we use the same initial robot positions and velocities as in Fig. 7.4b. Initially, the convex hull is maintained through a balance between the attraction forces from target-convergence constraints (7.19b) and the repulsion forces from neighboring collision-avoidance constraints (7.19c). When some robots break down and stop, as depicted in Fig. 7.8a, b, the remaining robots adjust by deviating away from the malfunctioning robots. They continue interacting with the operational neighbors, leading to the formation of quadrilateral and triangular patterns with different ordering sequences.

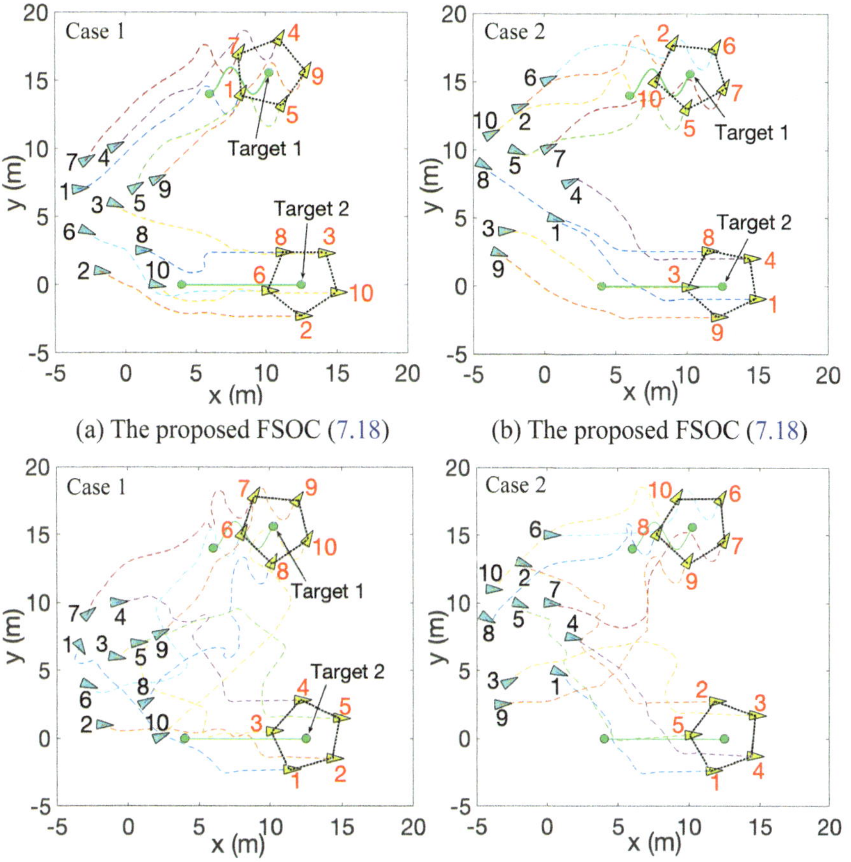

(a) The proposed FSOC (7.18) (b) The proposed FSOC (7.18)

(c) Fixed Subgroup + Convoying [17] (d) Fixed Subgroup + Convoying [17]

Fig. 7.4 First simulation: flexibility of the FSOC algorithm (7.19) **to different initial positions and comparison with fixed subgrouping and fencing algorithm** [19]. **a**, **b** Trajectories of ten robots from different initial positions to flexible fencing of two scattered targets with different subgroups and orderings. **c**, **d** Comparison trajectories under fixed subgrouping + fencing algorithm [19]. Here, the blue and yellow triangles denote the initial and final positions of robots, respectively; the dashed lines are the trajectories. The green circles denote two targets

In the second scenario, we simulate the appearance of a new target. In Fig. 7.9a, b, when a new target emerges at the bottom or top at $t \geq 3$ s, the ten robots initially form two subgroups to create pentagon formations around the original targets. Subsequently, they reconfigure into three subgroups to incorporate and successfully fence the new target, which demonstrates the algorithm's ability to adapt to new targets while maintaining the overall fencing strategy.

To evaluate the adaptability of the proposed FSOC algorithm (7.19) in environments with obstacles, we explore how different obstacle configurations affect the robots' behavior in the fourth set of simulations. We define and apply obstacle-

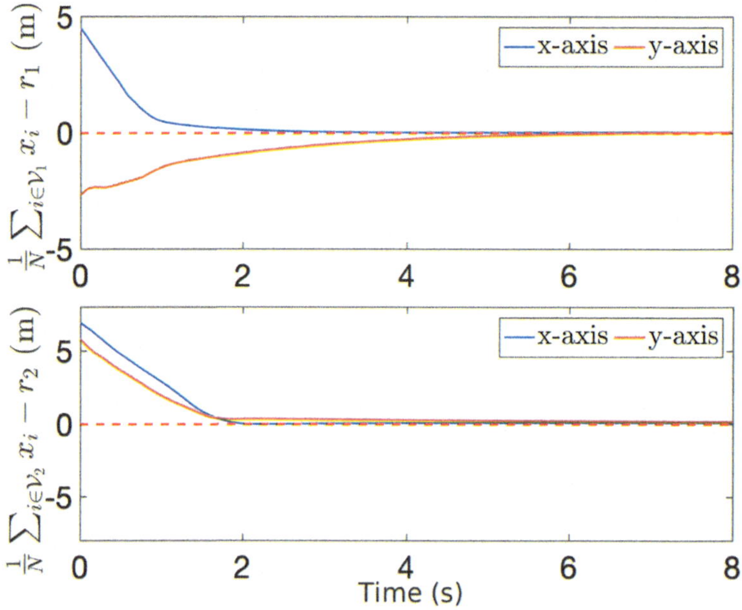

Fig. 7.5 Temporal evolution of the errors of flexible-ordering fencing $1/N \sum_{i\in\mathbb{V}_k} x_i - r_k, k = 1, 2$ in Fig. 7.4a for example

avoidance constraints similarly to those in Eqs. (7.15) and (7.16). In Fig. 7.10, as obstacles are introduced into the environment, robots must navigate around them, which can increase their travel distances. In Fig. 7.10a, b, we see that ten robots, starting from the same initial positions (marked as blue triangles), adjust their formations into three subgroups to efficiently circumvent obstacles. This reconfiguration not only demonstrates the algorithm's flexibility but also its efficiency in energy conservation. Additionally, Fig. 7.10c, which features a more densely packed obstacle field, shows the FSOC algorithm's ability to adaptively respond to complex and challenging environments.

To demonstrate the high-dimensional scalability of the proposed FSOC algorithm (7.19), we perform MST fencing simulations in a 3D environment. In Fig. 7.11a, we observe that targets 1 and 2 are fenced by five and four robots, respectively, with ordering sequences $\{7, 3, 2, 5, 9\}$ and $\{1, 8, 4, 6\}$. These sequences differ from the subgroups and ordering sequences $\{2, 8, 1, 4\}$ and $\{5, 6, 3, 7, 9\}$ shown in Fig. 7.11b, which confirms the scalability of the FSOC algorithm in 3D environments. Additionally, Fig. 7.11c incorporates scenarios with robot breakdowns and obstacles, further validating the algorithm's robustness and adaptability in complex settings. Thus, the FSOC algorithm's ability to handle high-dimensional and dynamic environments is effectively demonstrated.

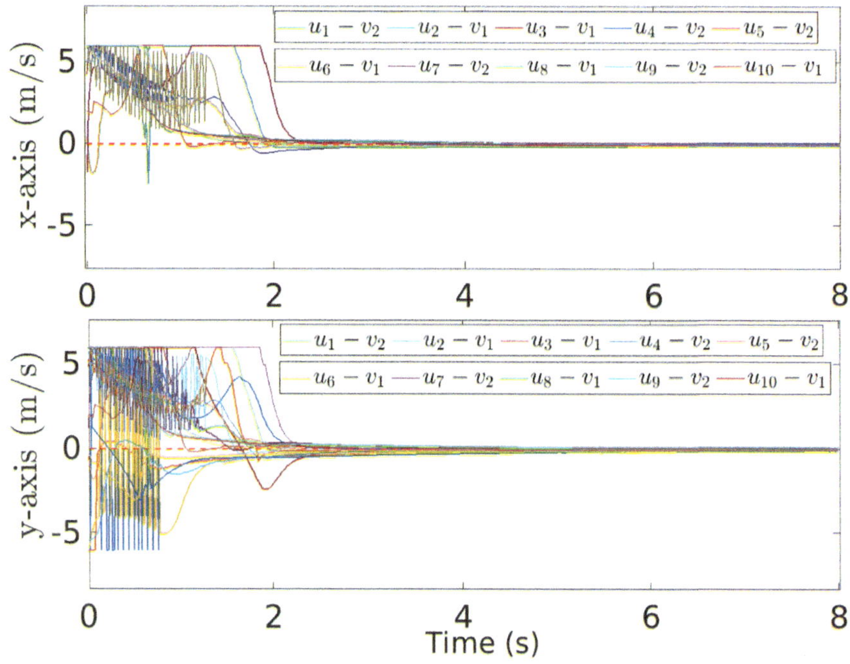

Fig. 7.6 Temporal evolution of the velocity-tracking errors $u_i - v_k$, $\forall i \in \mathbb{V}_k$, in Fig. 7.4a for example

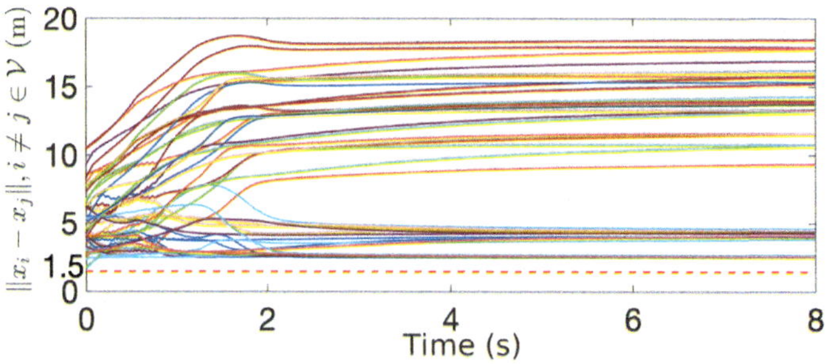

Fig. 7.7 Temporal evolution of the inter-robot distance $\|x_i - x_j\|$, $i \neq j \in \mathcal{V}$, in Fig. 7.4a for example

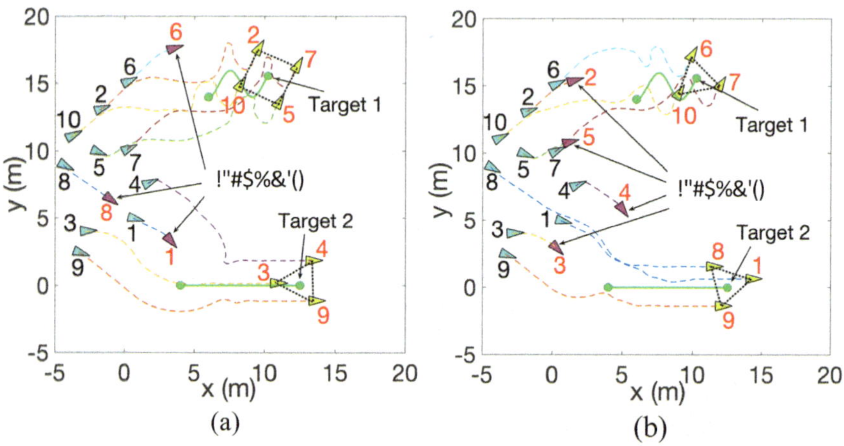

Fig. 7.8 Second simulation: robustness to robots suddenly breaking down. a Robots $i = 1, 6, 8$ break down at $T = 1$ s. **b** Robots $i = 2, 3, 4, 5$ break down at $T = 1$ s. (The blue, yellow triangles, green circles, and the dashed and solid green lines have the same meaning as in Fig. 7.4. The magenta triangles denote the final positions of the broken robots)

7.7 Conclusion

In this chapter, we have developed the FSOC algorithm, which allows multiple robots to dynamically organize into distinct subgroups and fence multiple scattered targets within their respective convex hulls, accommodating flexible ordering sequences. We have established conditions to ensure that the closed-loop multi-robot system asymptotically converges when using the FSOC algorithm. The algorithm's flexibility, adaptability, robustness, and scalability have been demonstrated through comprehensive 2D and 3D simulations. Future work will explore extending the FSOC algorithm to handle vehicles with general affine dynamics, manage polygonal obstacles, and address complex dynamics of targets.

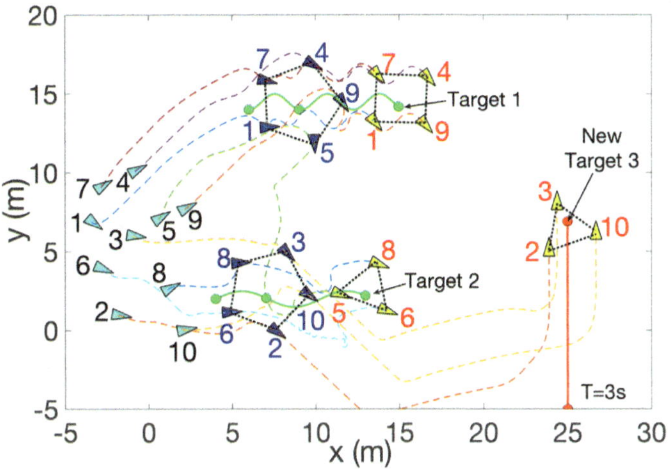

(a)

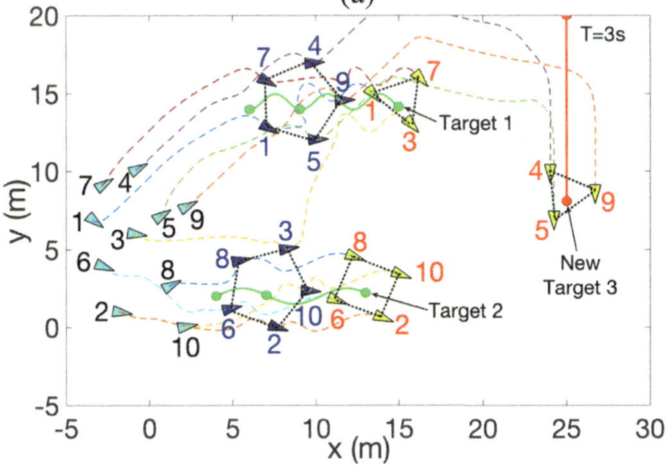

(b)

Fig. 7.9 Third simulation: robustness to a new target suddenly appearing. a The target 3 appears at the bottom with a velocities $v_3 = [0, 2]$ m/s at $T = 3$. **b** The target 3 appears at the top with a velocities $v_3 = [0, -2]$ m/s at $T = 3$. (The blue, yellow triangles, green circles, and the dashed and solid green lines have the same meaning as in Fig. 7.4. The dark blue triangles denote the positions of robots at $T = 3$ s)

Fig. 7.10 Four simulation: adaptability in obstacle environments. a, b Same initial positions of robots, same positions, and velocities of three scattered targets with different two obstacles result in the flexible fencing formations of different subgroups and ordering sequences. **c** Trajectories of ten robots from the same initial positions in Fig. 7.4a to the flexible fencing formations with different ordering sequences. (Most of the symbols have the same meaning as in Fig. 7.4. The red disks denote the obstacles. The purple numbers denote the different ordering sequences compared with Fig. 7.4)

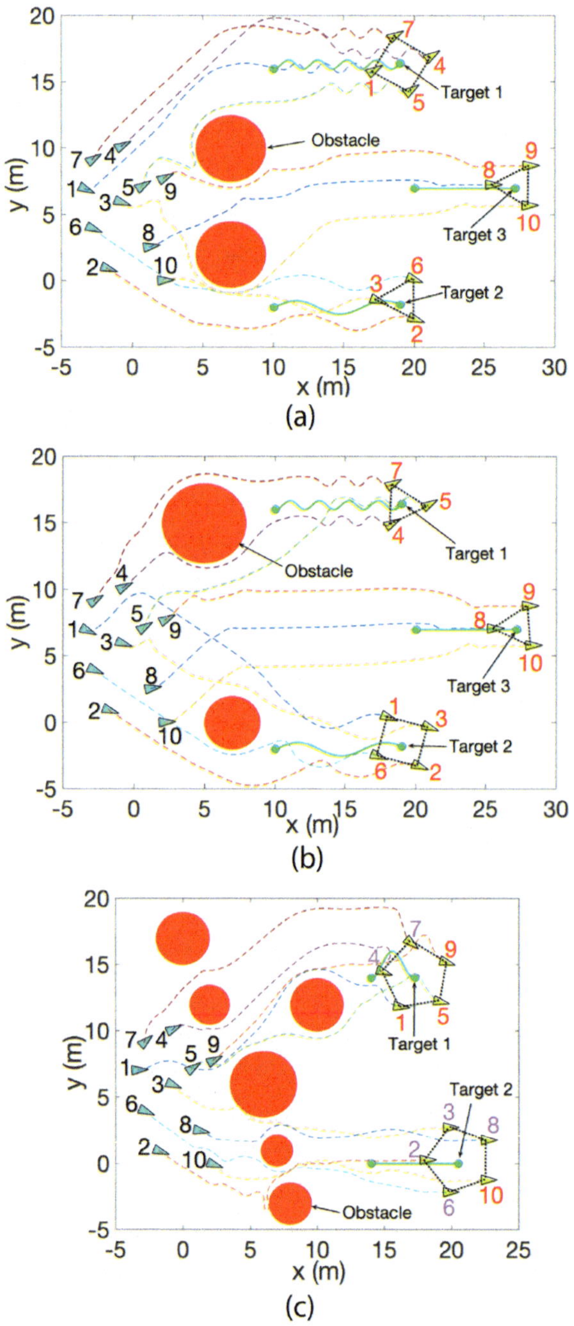

Fig. 7.11 Fifth simulation: scalability in 3D. a, b Trajectories of nine robots from different initial positions to 3D flexible fencing of two scattered targets with different subgroups and orderings. **c** A special situation when robot 2 suddenly breaks down at t = 1 s, and the rest of the eight robots from the same initial positions in Fig. 8a can spontaneously form regular tetrahedron fencing formations in the obstacle environments. (Most of the symbols have the same meaning as in Fig. 7.4. The dark blue balls denote the obstacles in 3D. The magenta triangle denotes that robot 2 is broken at $t = 1$ s)

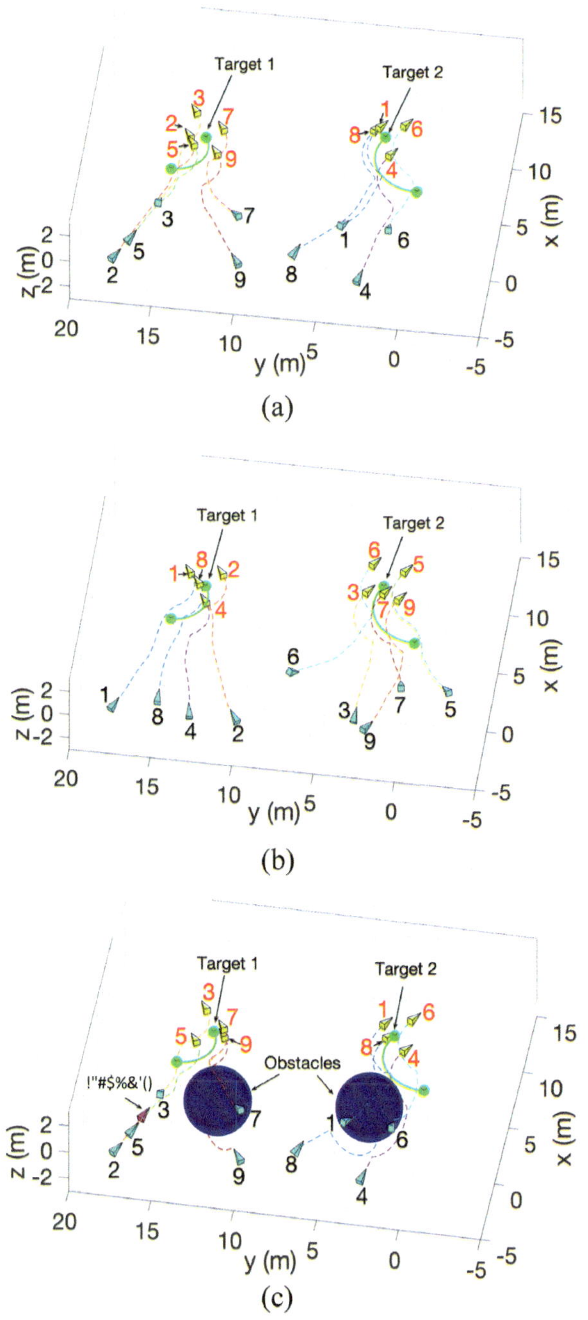

References

1. B.-B. Hu, H.-T. Zhang, B. Liu, J. Ding, Y. Xu, C. Luo, H. Cao, Coordinated navigation control of cross-domain unmanned systems via guiding vector fields. IEEE Trans. Control Syst. Technol. **32**(2), 550–563 (2024)
2. Z. Chen, Synchronization of frequency modulated multi-agent systems. IEEE Trans. Autom. Control **68**(6), 3425–3439 (2023)
3. H.-T. Zhang, B.-B. Hu, Z. Xu, Z. Cai, B. Liu, X. Wang, T. Geng, S. Zhong, J. Zhao, Visual navigation and landing control of an unmanned aerial vehicle on a moving autonomous surface vehicle via adaptive learning. IEEE Trans. Neural Netw. Learn. Syst. **32**(12), 5345–5355 (2021)
4. Z.-C. Xu, B.-B. Hu, B. Liu, X.-D. Wang, H.-T. Zhang, Vision-based autonomous landing of unmanned aerial vehicle on a motional unmanned surface vessel, in *Proceedings of the 39th Chinese Control Conference (CCC)* (IEEE, 2020), pp. 6845–6850
5. F. Chen, W. Ren, Y. Cao, Surrounding control in cooperative agent networks. Syst. Control Lett. **59**(11), 704–712 (2010)
6. W. Meng, Z. He, R. Su, P.K. Yadav, R. Teo, L. Xie, Decentralized multi-UAV flight autonomy for moving convoys search and track. IEEE Trans. Control Syst. Technol. **25**(4), 1480–1487 (2016)
7. B.-B. Hu, H.-T. Zhang, B. Liu, H. Meng, G. Chen, Distributed surrounding control of multiple unmanned surface vessels with varying interconnection topologies. IEEE Trans. Control Syst. Technol. **30**(1), 400–407 (2021)
8. Y.J. Shi, R. Li, K.L. Teo, Cooperative enclosing control for multiple moving targets by a group of agents. Int. J. Control **88**(1), 80–89 (2015)
9. B.-B. Hu, H.-T. Zhang, J. Wang, Multiple-target surrounding and collision avoidance with second-order nonlinear multiagent systems. IEEE Trans. Ind. Electron. **68**(8), 7454–7463 (2020)
10. B. Xu, H.-T. Zhang, Y. Ding, W. Ren, Event-triggered surrounding formation control of multi-agent systems for multiple dynamic targets. IEEE Trans. Control Netw. Syst. **10**(2), 752–764 (2020)
11. K. Chen, G. Qi, Y. Li, A. Sheng, Cooperative localization and circumnavigation of multiple targets with bearing-only measurements. J. Frankl. Inst. **360**(12), 9159–9179 (2023)
12. Z. Sun, H.G. de Marina, G.S. Seyboth, B.D.O. Anderson, C. Yu, Circular formation control of multiple unicycle-type agents with nonidentical constant speeds. IEEE Trans. Control Syst. Technol. **27**(1), 192–205 (2018)
13. L. Shi, R. Zheng, M. Liu, S. Zhang, Distributed control of cooperative multi-target enclosing by ring-coupled unicycles, in *Proceedings of the American Control Conference (ACC)*, pp. 3523–3528 (2020)
14. K. Sakurama, H.-S. Ahn, Multi-agent coordination over local indexes via clique-based distributed assignment. Automatica **112**, 108670 (2020)
15. Z. Chen, A cooperative target-fencing protocol of multiple vehicles. Automatica **107**, 591–594 (2019)
16. B. Liu, Z. Chen, H.-T. Zhang, X. Wang, T. Geng, H. Su, J. Zhao, Collective dynamics and control for multiple unmanned surface vessels. IEEE Trans. Control Syst. Technol. **28**(6), 2540–2547 (2019)
17. L. Kou, Z. Chen, J. Xiang, Cooperative fencing control of multiple vehicles for a moving target with an unknown velocity. IEEE Trans. Autom. Control **67**(2), 1008–1015 (2021)
18. L. Dou, X. Yu, L. Liu, X. Wang, G. Feng, Moving-target enclosing control for mobile agents with collision avoidance. IEEE Trans. Control Netw. Syst. **8**(4), 1669–1679 (2021)
19. B.-B. Hu, H.-T. Zhang, Y. Shi, Cooperative label-free moving target fencing for second-order multi-agent systems with rigid formation. Automatica **148**, 110788 (2023)
20. M. Turpin, N. Michael, V. Kumar, CAPT: concurrent assignment and planning of trajectories for multiple robots. Int. J. Robot. Res. **33**(1), 98–112 (2014)
21. J. Alonso-Mora, A. Breitenmoser, M. Rufli, R. Siegwart, P. Beardsley, CAPT: image and animation display with multiple mobile robots. Int. J. Robot. Res. **33**(6), 753–773 (2012)

22. G. Notomista, M. Egerstedt, Constraint-driven coordinated control of multi-robot systems, in *Proceedings of the American Control Conference (ACC)*, pp. 1990–1996 (2019)
23. B.-B. Hu, H.-T. Zhang, W. Yao, J. Ding, M. Cao, Spontaneous-ordering platoon control for multirobot path navigation using guiding vector fields. IEEE Trans. Robot. **39**(4), 2654–2668 (2023)
24. A.D. Ames, X. Xu, J.W. Grizzle, P. Tabuada, Control barrier function based quadratic programs for safety-critical systems. IEEE Trans. Autom. Control **62**(8), 3861–3876 (2016)
25. G. Notomista, S. Mayya, Y. Emam, C. Kroninger, A. Bohannon, S. Hutchinson, M. Egerstedt, A resilient and energy-aware task allocation framework for heterogeneous multirobot systems. IEEE Trans. Robotics **38**(1), 159–179 (2021)
26. S. Boyd, L. Vandenberghe, *Convex Optimization* (Cambridge University Press, UK, 2004)
27. N.J. Higham, *Accuracy and Stability of Numerical Algorithms* (SIAM, UK, 2004)
28. K.H. Khalil, *Nonlinear Systems* (Prentice Hall, Upper Saddle River, 2002)
29. P. Glotfelter, I. Buckley, M. Egerstedt, Hybrid nonsmooth barrier functions with applications to provably safe and composable collision avoidance for robotic systems. IEEE Robot. Autom. Lett. **4**(2), 1303–1310 (2019)

Part III
Bearing-Only Scenario

Chapter 8
Bearing-Only Fencing via Persistent-Exciting Condition

8.1 Introduction

In the previous Chaps. 2–7, we have conducted detailed research on two themes for the cooperative control of multi-robot systems, namely, "basic fencing" and "complex fencing." However, the cooperative methods in these studies rely on the relative position between the robot and the target. However, such kind of information usually depends on distance or position sensors in practice which are costly and challenging to implement on a large-scale system. Considering the usage of simple and cheap orientation sensors can significantly reduce costs and is of great importance for large-scale scenarios. However, the incompleteness and constraints of orientation information lead to issues in distributed cooperative fencing. Therefore, based on the previous "basic fencing" and "complex fencing," this chapter proposes a distributed cooperative bearing-only fencing control algorithm via the persistent-exciting condition, which aims to delve into the distributed cooperative fencing problem. In what follows, we will firstly introduce some related works.

Over the years, the control of USVs has garnered increasing attention due to their effectiveness, versatility, and ability to adapt to complex environments, which has been extended to applications such as trajectory tracking [1–5], path following [6–9], and collective exploration [10–14]. Cooperative target-fencing missions using multiple vessels are particularly important for tasks such as surveillance, detection, and escort formations. Therein, the study of USV target-fencing problems originated from multi-agent systems (MASs), where theoretical challenges initially focused efforts on strategies involving both position and velocity measurements. Chen and Ren et al. [15] developed an estimation-and-control framework with a fixed topology for surrounding stationary targets. Lan et al. [16] proposed a hybrid control strategy for static target fencing. Later, Liu and Chen et al. [17] introduced a hierarchical output regulation scheme for target fencing, while Chen [18] offered a cooperative control approach incorporating attractive, repulsive, and rotational components to handle specific target-fencing problems with assured collision avoidance.

B.-B. Hu and H.-T. Zhang, *Multi-Robot Systems*, Cognitive Intelligence and Robotics,
https://doi.org/10.1007/978-981-97-9214-6_8

As aquatic missions become increasingly complex, more focus has been placed on encircling moving targets. For instance, Guo et al. [19] proposed a distributed protocol that relies only on neighboring positions to fence a moving target. Yu and Liu developed a distributed dynamic controller for surrounding a moving target under both fixed [20] and switching [21] inter-agent topologies. Shoja et al. [22] later introduced an adaptive method to handle the fencing of multiple moving targets by non-identical agents. However, the above studies [19–22] mainly addressed first-order MASs due to the significant difficulties associated with fencing control in higher-order MASs. To address more practical scenarios, Shi et al. [23] investigated a second-order MAS fencing problem using Lasalle's Invariance Principle [24]. Li et al. [25] introduced a cooperative fencing strategy that ensures collision avoidance for networked MASs with nonlinear Lagrangian dynamics. Hu et al. [26] designed a distributed equal-distance fencing controller for second-order nonlinear MASs. Peng et al. [27] later presented an event-triggered dynamic surface controller for a USV to encircle a dynamic target. Xu [28] proposed a novel fault-tolerant control scheme for a fleet of USVs using only line-of-sight range and angle to track a target. Hu et al. [29] developed a distributed fencing controller for multiple USVs with varying interconnection topologies. Kou et al. [30] studied the problem of multiple second-order vehicles fencing a moving target with an unknown constant velocity.

However, the high cost of onboard sensors makes it challenging for USVs to consistently obtain both the position and velocity of targets in real-world scenarios. This challenge has prompted research into approaches that rely solely on bearing measurements (e.g., bearing vectors or subtended bearing angles), which only require more affordable sensors like pinhole cameras and wireless sensor arrays. Among the pioneering works in this area, Zheng et al. [31] introduced a control strategy that depends purely on local bearing measurements to achieve the enclosure of a stationary target. Li et al. [32] developed a distributed bearing-only controller for targeting stationary objects in 3D environments. Deghat et al. [33] investigated the localization and circumnavigation of a moving target with unknown velocity. Yang et al. [34] explored a target-entrapping problem using only bearing data, while Dou et al. [35] proposed a distributed controller that enables agents to encircle a moving target and form desired collective patterns based exclusively on bearing measurements.

Despite these advancements, the above-mentioned bearing-based methods [31–35] primarily focus on first-order MASs and do not account for more complex USV scenarios with significant nonlinearity and strong coupling dynamics. Although position-based target-fencing strategies for USVs have been studied in [17, 26, 27, 29], addressing the theoretical challenges of controlling USV groups using only bearing measurements remains an unresolved issue [33–35].

From a technical standpoint, the key difficulties lie in the presence of bounded errors and the limited availability of target velocity information to the USVs, which hinders the broader implementation of multi-USV cooperative control strategies. In response, this chapter introduces a tailored distributed protocol for multi-USV systems that effectively reduces fencing errors to zero using only bearing feedback. Beyond enabling the fencing of a moving target, the proposed controller ensures the USVs continuously rotate around the target vessel with dynamically changing topologies, essential for the asymptotic convergence of bearing-only target estimation. In contrast to most existing studies that rely solely on simulations [32–35], this work includes experiments conducted on a real-world multi-USV platform consisting of a USV motion capture system, a monitoring station, three 3D-printed HUSTER-0.3 vessels, and a target vessel, thereby validating the proposed framework. The contribution of the chapter is two-fold.

1. Propose a distributed bearing-only controller for the multi-USV systems to collectively fence and rotate evenly around a motional target vessel, with the design of a bearing-only target estimator, an upper-level fencing protocol, and a single vessel regulator.
2. Conduct indoor experiments with three HUSTER-0.3 USVs to verify the effectiveness of the proposed algorithm.

The rest of this chapter is structured as follows. Section 8.2 presents the problem formulation. Section 8.3 details the main technical results, along with comprehensive mathematical proofs. In Sect. 8.4, experimental validation is carried out on a real multi-USV platform. Finally, Sect. 8.5 concludes the chapter.

8.2 Problem Formulation of Bearing-Only Target Fencing

In this section, we firstly consider a multi-USV system with n USVs, each of which is characterized by its kinematic equations in Cartesian coordinates, as detailed in [36].

$$\begin{aligned} \dot{x}_i &= u_i \cos\psi_i - v_i \sin\psi_i, \\ \dot{y}_i &= u_i \sin\psi_i + v_i \cos\psi_i, \\ \dot{\psi}_i &= r_i, \end{aligned} \tag{8.1}$$

where the position of USV i is represented as $q_i(t) = [x_i(t), y_i(t)]^\mathsf{T} \in \mathbb{R}^2$, while $\psi_i(t) \in \mathbb{R}$ denotes its yaw angle. Additionally, $u_i(t)$, $v_i(t)$, and $r_i(t) \in \mathbb{R}$ correspond to the surge, sway, and yaw velocities of USV i in its own coordinate frame, respectively.

Moreover, The dynamics of USV i typically adhere to a practical model (see, e.g., [17]) given as

$$\begin{aligned} \dot{u}_i &= k_1 u_i + k_2 v_i r_i + k_3 \tau_{i,1}, \\ \dot{r}_i &= k_4 r_i + k_5 \tau_{i,2}, \\ \dot{v}_i &= k_6 v_i + k_7 u_i r_i. \end{aligned} \tag{8.2}$$

Here, $k_1, k_2, k_3, k_4, k_5, k_6, k_7 \in \mathbb{R}$ are parameters determined via zig-zag tests, while $\tau_{i,1}$ and $\tau_{i,2} \in \mathbb{R}$ denote the control inputs for USV i.

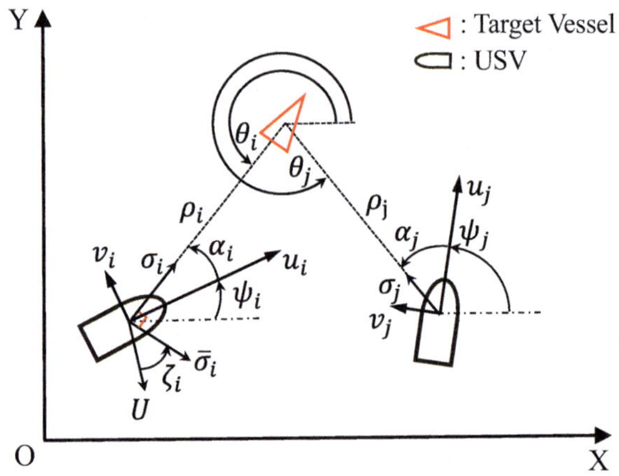

Fig. 8.1 Illustration of the bearing angle in the multi-USV system

Secondly, we consider a motional target to be $q_d(t) = [x_d(t), y_d(t)]^{\mathrm{T}} \in \mathbb{R}^2$ satisfies

$$\dot{q}_d = p_d \tag{8.3}$$

where $p_d = [v_x^d, v_y^d]^{\mathrm{T}} \in \mathbb{R}^2$ is a constant velocity.

As illustrated in Fig. 8.1, it is assumed that only the relative bearing angle α_i is measured between the ith USV and the target within the ith USV's coordinate frame, which is not necessarily aligned with the coordinate frames of the other USVs. The position q_d and the velocity p_d of the target vessel are not directly measured. Considering the yaw angle ψ_i, one obtains

$$\sigma_i := \frac{q_{d,i}}{\|q_{d,i}\|} \tag{8.4}$$

where $\angle\sigma_i := \psi_i + \alpha_i$ and $q_{d,i} := q_d - q_i$ in Eq. (8.18).

Note that the communication topology of the multi-USV system is typically characterized by $\mathcal{G}(\mathcal{V}, \mathcal{E})$, where $\mathcal{V} := \{1, 2, \ldots, n\}$ is the set of nodes, and $\mathcal{E} := \{(i, j), j \in \mathcal{N}_i, i \in \mathcal{V}\}$ represents the set of edges. The neighbor set of USV i is defined as $\mathcal{N}_i = \{j \in \mathcal{V}|\ \min_{j\neq i}\Delta_{i,j}, \min_{j\neq i}\Delta_{j,i}\}$, where $\Delta_{i,j} \in [0, 2\pi), i \neq j \in \mathcal{V}$ is the forward separation angle from USV i to USV j, obtained by rotating σ_i counterclockwise until it coincides with σ_j. Conversely, $\Delta_{j,i} \in [0, 2\pi)$ denotes the backward separation angle from USV j to USV i, as illustrated in Fig. 8.2 (see, e.g., [35]). Note that $\mathcal{N}_i$ may encompass more than two neighbors, and both the pre-neighbor i^+ and the next-neighbor i^- of USV i can be randomly chosen from $\mathcal{N}_i$.

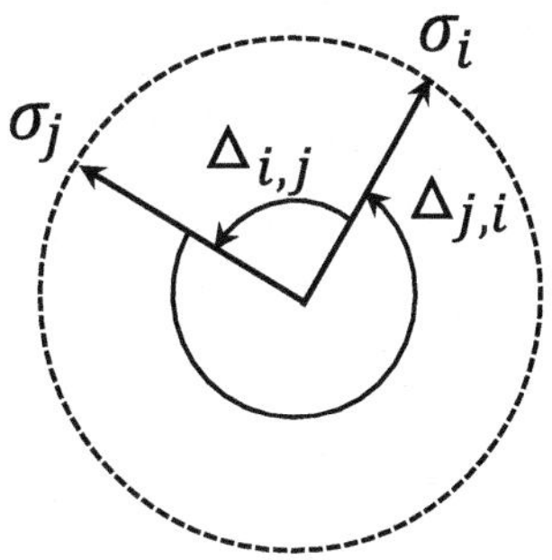

Fig. 8.2 Illustration of the forward and backward separation angles $\Delta_{i,j}$ and $\Delta_{j,i}$, respectively, between vector σ_i of USV i and vector σ_j of USV j

We define the relative distance ρ_i and the bearing angle θ_i between USV i and the target vessel to be

$$\rho_i = \sqrt{(x_i - x_d)^2 + (y_i - y_d)^2},$$
$$\theta_i = \text{arctan2}(y_i - y_d, x_i - x_d) + 2\kappa\pi \in [0, 2\pi) \tag{8.5}$$

where κ is an integer selected to ensure that $\theta \in [0, 2\pi)$. The function arctan2(y, x) computes the angle of the vector (x, y), returning a value between $-\pi$ and π radians. Consequently, one has

$$\theta_i = \psi_i + \alpha_i + \pi \tag{8.6}$$

where α_i and ψ_i are shown in Fig. 8.1. The separation angle ς_i between USV i and its pre-neighbor i^+, with respect to the target vessel, is defined as

$$\varsigma_i := \theta_{i^+} - \theta_i + \xi_i \tag{8.7}$$

with

$$\xi_i = \begin{cases} 0, & \theta_{i^+} - \theta_i \geq 0, \\ 2\pi, & \theta_{i^+} - \theta_i < 0. \end{cases}$$

Based on the target and neighboring information provided, the following assumptions are needed for the subsequent analysis.

Assumption 8.1 The target's velocity $\dot{p}_d \in \mathbb{R}^2$ in (8.3) is constrained by an upper bound $p_d^* \in \mathbb{R}^+$, such that $\|\dot{p}_d\| \leq p_d^*$.

Assumption 8.1 is valid because, in practice, the target vessel's energy is inherently finite, which means that there is an upper limit on the target's velocity.

Assumption 8.2 For each USV $i \in \mathcal{V}$, the communication neighbor set $\mathcal{N}_i^c = \{i^+, i^-\}$.

Assumption 8.2 illustrates that each USV is capable of communicating with its pre-neighbor i^+ and next-neighbor i^-. The communication topology $\mathcal{G}(\mathcal{V}, \mathcal{E})$, which is derived from the minimal-neighbor strategy, maintains strong connectivity even when neighbors change over time [37]. One will have a cost-effective communication topology, which is beneficial for real-world applications. To proceed, it is also essential to define target fencing.

Definition 8.1 (*Target fencing*) Target fencing by the USVs is achieved asymptotically with the following characteristics:

$$\lim_{t\to\infty} \rho_i(t) = \rho_d, \lim_{t\to\infty} \varsigma_i = \frac{2\pi}{n}, \tag{8.8}$$

where $\rho_d \in \mathbb{R}^+$ denotes the desired radius, and ς_i represents the separation angle as given in (8.7).

Definition 8.1 demonstrates that the USVs evenly surround the target vessel, meaning that the target vessel lies within the convex hull created by the USVs [17, 18]. Thus, one has that

$$\lim_{t\to\infty} P_{x_d(t)}(x(t)) = 0$$

where $x := [x_1^\top, \ldots, x_n^\top]^\top$, $P_{x_d}(x) := \min_{s\in\text{co}(x)} \|x_d - s\|$ and

$$\text{co}(x) := \left\{ \sum_{i\in\mathcal{V}} \lambda_i x_i : \lambda_i \geq 0, \forall i \text{ and } \sum_{i\in\mathcal{V}} \lambda_i = 1 \right\}.$$

Taking the temporal derivative of ρ_i along the dynamics of Eq. (8.5) yields

$$\begin{aligned}\dot{\rho}_i &= \frac{(x_i - x_d)(\dot{x}_i - v_x^d) + (y_i - y_d)(\dot{y}_i - v_y^d)}{\sqrt{(x_i - x_d)^2 + (y_i - y_d)^2}} \\ &= \begin{bmatrix}\cos\theta_i & \sin\theta_i\end{bmatrix} \begin{bmatrix}\dot{x}_i - v_x^d \\ \dot{y}_i - v_y^d\end{bmatrix}.\end{aligned} \tag{8.9}$$

Analogously, one has that the temporal derivative of θ_i is

$$\begin{aligned}\dot{\theta}_i &= \frac{(x_i - x_d)^2}{\rho_i^2} \frac{(x_i - x_d)(\dot{y}_i - v_y^d) - (y_i - y_d)(\dot{x}_i - v_x^d)}{(x_i - x_d)^2} \\ &= \frac{\begin{bmatrix}-\sin\theta_i & \cos\theta_i\end{bmatrix}}{\rho_i} \begin{bmatrix}\dot{x}_i - v_x^d \\ \dot{y}_i - v_y^d\end{bmatrix}.\end{aligned} \tag{8.10}$$

Rewriting Eqs. (8.9) and (8.10) into a compact form yields

$$\begin{bmatrix}\dot{\rho}_i \\ \rho_i\dot{\theta}_i\end{bmatrix} = \begin{bmatrix}\cos\theta_i & \sin\theta_i \\ -\sin\theta_i & \cos\theta_i\end{bmatrix} \begin{bmatrix}\dot{x}_i - v_x^d \\ \dot{y}_i - v_y^d\end{bmatrix}, \tag{8.11}$$

which implies

$$\begin{bmatrix} \dot{x}_i \\ \dot{y}_i \end{bmatrix} = \begin{bmatrix} \cos(-\theta_i) & \sin(-\theta_i) \\ -\sin(-\theta_i) & \cos(-\theta_i) \end{bmatrix} \begin{bmatrix} \dot{\rho}_i \\ \rho_i \dot{\theta}_i \end{bmatrix} + \begin{bmatrix} v_x^d \\ v_y^d \end{bmatrix}. \tag{8.12}$$

Together with Eq. (8.1), one has

$$\begin{bmatrix} u_i \\ v_i \end{bmatrix} = \begin{bmatrix} \cos\psi_i & \sin\psi_i \\ -\sin\psi_i & \cos\psi_i \end{bmatrix} \begin{bmatrix} \dot{x}_i \\ \dot{y}_i \end{bmatrix}, \tag{8.13}$$

which follows from Eq. (8.13) and (8.12) that

$$\begin{bmatrix} u_i \\ v_i \end{bmatrix} = \begin{bmatrix} \cos(\psi_i - \theta_i) & \sin(\psi_i - \theta_i) \\ -\sin(\psi_i - \theta_i) & \cos(\psi_i - \theta_i) \end{bmatrix} \begin{bmatrix} \dot{\rho}_i \\ \rho_i \dot{\theta}_i \end{bmatrix} + \begin{bmatrix} \cos\psi_i & \sin\psi_i \\ -\sin\psi_i & \cos\psi_i \end{bmatrix} \begin{bmatrix} v_x^d \\ v_y^d \end{bmatrix}. \tag{8.14}$$

From the condition that $\psi_i - \theta_i = -\alpha_i - \pi$ in (8.6), it can be deduced that

$$\begin{bmatrix} u_i \\ v_i \end{bmatrix} = \begin{bmatrix} -\cos\alpha_i & \sin\alpha_i \\ -\sin\alpha_i & -\cos\alpha_i \end{bmatrix} \begin{bmatrix} \dot{\rho}_i \\ \rho_i \dot{\theta}_i \end{bmatrix} + \begin{bmatrix} \cos\psi_i & \sin\psi_i \\ -\sin\psi_i & \cos\psi_i \end{bmatrix} \begin{bmatrix} v_x^d \\ v_y^d \end{bmatrix}. \tag{8.15}$$

Define u_i^r, v_i^r as the upper-level signals for u_i, v_i, respectively, in the kinematics (8.1), we can further define the signal errors to be

$$\widetilde{u}_i := u_i - u_i^r, \widetilde{v}_i := v_i - v_i^r, \tag{8.16}$$

it, together with Eqs. (8.15) and (8.16), gives that the dynamics of ρ_i, θ_i is formulated to be

$$\begin{bmatrix} \dot{\rho}_i \\ \rho_i \dot{\theta}_i \end{bmatrix} = \begin{bmatrix} -\cos\alpha_i & \sin\alpha_i \\ -\sin\alpha_i & -\cos\alpha_i \end{bmatrix} \begin{bmatrix} u_i^r \\ v_i^r \end{bmatrix} + \begin{bmatrix} \bar{u}_i \\ \bar{v}_i \end{bmatrix} - \begin{bmatrix} \cos\theta_i & \sin\theta_i \\ -\sin\theta_i & \cos\theta_i \end{bmatrix} \begin{bmatrix} v_x^d \\ v_y^d \end{bmatrix} \tag{8.17}$$

where $\bar{u}_i := -\cos\alpha_i \widetilde{u}_i + \sin\alpha_i \widetilde{v}_i$ and $\bar{v}_i := -\sin\alpha_i \widetilde{u}_i - \cos\alpha_i \widetilde{v}_i$. We are now prepared to present the two main problems addressed in this chapter.

Problem 8.1 (*Target fencing*) Design an upper-level fencing protocol

$$u_i^r, v_i^r := f(\alpha_i, \theta_i, \theta_j), \quad j \in \mathcal{N}_i$$

for the multi-USV system described by (8.1), (8.3), and (8.17) to achieve the motional target-fencing (see Definition 8.1), while ensuring that $\widetilde{u}_i$ and $\widetilde{v}_i$ approach zero.

Problem 8.2 (*Vessel Regulation*) Design an actuator input

$$\tau_1, \tau_2 := f(u^r, v^r, u, v, r)$$

for a single vessel, as described by (8.2) without the subscript i, such that the velocities u and v converge to the upper-level signals u^r and v^r from Problem 8.1, i.e., ensuring $\lim_{t\to\infty}\widetilde{u}(t)=0$ and $\lim_{t\to\infty}\widetilde{v}(t)=0$.

8.3 Perturbed-Stable Controller and Convergence Analysis

8.3.1 Target Fencing Control

We define $q_{d,i}$ to be the relative position between the USV i and the target vessel below

$$q_{d,i} := q_d - q_i \tag{8.18}$$

where q_i, q_d are given in Eqs. (8.1) and (8.3), respectively.

Due to the fact that the position $q_{d,i}$ and velocity p_d of the target are not available, we define $\widehat{q}_{d,i}$ and $\widehat{p}_{d,i}=[\widehat{v}_x,\widehat{v}_y]^\top$ as the estimates of $q_{d,i}$ and p_d, respectively, and design the estimates using the bearing-only measurements

$$\begin{aligned}\dot{\widehat{q}}_{d,i}(t)=&-\dot{q}_i(t)-c_1(I-\sigma_i(t)\sigma_i(t)^\top)\widehat{q}_{d,i}(t)-c_2\int_0^t(I-\sigma_i(\epsilon)\sigma_i(\epsilon)^\top)\widehat{q}_{d,i}(\epsilon)d\epsilon,\\ \dot{\widehat{p}}_{d,i}(t)=&-c_2(I-\sigma_i(t)\sigma_i(t)^\top)\widehat{q}_{d,i}(t)dt,\end{aligned} \tag{8.19}$$

with $c_1, c_2\in\mathbb{R}^+$ being the estimated gains and σ_i defined Eq. (8.4). Then, we propose a distributed bearing-only controller,

$$\begin{bmatrix}u_i^r\\ v_i^r\end{bmatrix}=\begin{bmatrix}-\cos\alpha_i & -\sin\alpha_i\\ \sin\alpha_i & -\cos\alpha_i\end{bmatrix}\begin{bmatrix}c_3(\rho_d-\widehat{\rho}_i)+\cos\theta_i\widehat{v}_x+\sin\theta_i\widehat{v}_y\\ c_4(\varsigma_i-\varsigma_{i^-})+\beta-\sin\theta_i\widehat{v}_x+\cos\theta_i\widehat{v}_y\end{bmatrix}, \tag{8.20}$$

where $c_3, c_4\in\mathbb{R}^+$ represent the control gains, ρ_d denotes the desired radius in Definition 8.1, $\widehat{\rho}_i := \|\widehat{q}_{d,i}\|$ is the estimated radius in Eq. (8.19), and ς_i, ς_{i^-} are the separation angles between the unmanned surface vehicle (USV) i and its next-neighbor i^-, respectively.

Remark 8.1 In contrast to target information that is available only to a subset of agents (e.g., position-based target-fencing controllers [20–22]), each USV in this chapter has access to the bearing information σ_i between itself and the target vessel, where such kind of bearing is an integral part of the bearing-only target estimator (8.19). The key reason is that σ_i is measured in each USV's coordinate system and cannot be directly shared among USVs, where this assumption is widely accepted in the literature on bearing-only methods (see, e.g., [33–35]).

To lay the groundwork for the main technical result, we first need to introduce the following lemmas.

Lemma 8.1 ([35]) *Consider a perturbed system*

$$\dot{x} = A(t, d(t))x + h(t), \tag{8.21}$$

where $x \in \mathbb{R}^n$, $h(t) \in \mathbb{R}^n$, $d(t) \in \mathcal{D}$ *denotes a time-varying function, and* $A : \mathbb{R}^+ \times \mathcal{D} \to \mathbb{R}^{n\times n}$ *is a continuous and globally bounded function. If the nominal system* $\dot{x} = A(t, d(t))x$ *is uniformly globally exponentially stable and* $h(t)$ *is bounded with* $\lim_{t\to\infty} h(t) = 0$, *one has that* $\lim_{t\to\infty} x(t) = 0$.

Lemma 8.2 ([37]) *For a connected and symmetric matrix* $L \in \mathbb{R}^{n\times n}$, *the following holds:*

$$\min_{\mathbf{1}^\top x=0, x\neq 0} \frac{x^\top L x}{\|x\|^2} \geq \lambda_2(L) > 0. \tag{8.22}$$

Lemma 8.3 ([35]) *Consider a nominal system*

$$\dot{\widetilde{q}}_{d,i} = -\bar{\sigma}_i \bar{\sigma}_i^\top \widetilde{q}_{d,i}, \tag{8.23}$$

where $\widetilde{q}_{d,i} := \widehat{q}_{d,i} - q_{d,i} \in \mathbb{R}^2$ *denotes the estimated error given in* (8.26), *and* $\bar{\sigma}_i$, $i \in \mathcal{V}$, *is a unit vector perpendicular to* σ_i *with respect to the angle* $\angle\bar{\sigma}_i = \psi_i + \alpha_i - \pi/2$ *in Fig. 8.1. Then, one has that*

$$\lim_{t\to\infty} \widetilde{q}_{d,i}(t) = 0,$$

i.e., $\bar{\sigma}_i(t)$, $i \in \mathcal{V}$, *is persistently exciting [38], if* β *in* (8.20) *fulfills the following condition*

$$\beta > 2\pi c_4 + p_d^* + \beta^* \tag{8.24}$$

where $\beta^* \in \mathbb{R}^+$ *is a constant and* $c_4 \in \mathbb{R}^+$ *denotes the control gain in* (8.20).

Lemma 8.4 *Consider the estimator of the state* $q_{d,i}$, p_d *given by* (8.1), (8.3), (8.18), (8.19), *it holds that*

$$\lim_{t\to\infty} \widehat{q}_{d,i}(t) = q_{d,i}, \lim_{t\to\infty} \widehat{p}_{d,i}(t) = p_d, \tag{8.25}$$

if β *in* (8.20) *satisfies* (8.24) *in Lemma 8.3.*

Proof We define $\widetilde{q}_{d,i}$, $\widetilde{p}_d$ to be the estimated errors,

$$\widetilde{q}_{d,i} := \widehat{q}_{d,i} - q_{d,i}, \ \widetilde{p}_{d,i} := \widehat{p}_{d,i} - p_d. \tag{8.26}$$

which implies that

$$\dot{\widetilde{q}}_{d,i} = \dot{\widehat{q}}_{d,i} - \dot{q}_{d,i}, \ \dot{\widetilde{p}}_{d,i} = \dot{\widehat{p}}_{d,i}. \tag{8.27}$$

It follows from Eqs. (8.3), (8.19) and (8.27) that

$$\begin{aligned}\dot{\tilde{q}}_{d,i} &= -c_1(I-\sigma_i\sigma_i^{\mathsf{T}})(\tilde{q}_{d,i}+q_{d,i})+\tilde{p}_{d,i},\\ \dot{\tilde{p}}_{d,i} &= -c_2(I-\sigma_i\sigma_i^{\mathsf{T}})(\tilde{q}_{d,i}+q_{d,i}).\end{aligned} \tag{8.28}$$

From the condition that

$$(I-\sigma_i\sigma_i^{\mathsf{T}})q_{d,i} = q_{d,i} - \frac{q_{d,i}}{\|q_{d,i}\|}\frac{q_{d,i}^{\mathsf{T}}}{\|q_{d,i}\|}q_{d,i} = 0$$

and

$$\bar{\sigma}_i\bar{\sigma}_i^{\mathsf{T}} = I-\sigma_i\sigma_i^{\mathsf{T}},$$

it can be deduced that

$$\begin{bmatrix}\dot{\tilde{q}}_{d,i}\\ \dot{\tilde{p}}_d\end{bmatrix} = A\begin{bmatrix}\tilde{q}_{d,i}\\ \tilde{p}_d\end{bmatrix} \tag{8.29}$$

with

$$A = \begin{bmatrix}-c_1\bar{\sigma}_i\bar{\sigma}_i^{\mathsf{T}} & I\\ -c_2\bar{\sigma}_i\bar{\sigma}_i^{\mathsf{T}} & 0\end{bmatrix}. \tag{8.30}$$

Here, $\bar{\sigma}_i$ represents the unit vector perpendicular to σ_i, with an angle $\angle\bar{\sigma}_i = \psi_i + \alpha_i - \frac{\pi}{2}$ as depicted in Fig. 8.1. The convergence of the error dynamics given by (8.29) is ensured if the matrix A is Hurwitz. This is equivalent to demonstrating that $\bar{\sigma}_i$ for $i \in \mathcal{V}$ is persistently exciting [38].

According to Lemma 8.3, it follows that $\bar{\sigma}_i$ is persistently exciting for a suitably chosen β in (8.24). Consequently, the estimated errors $\tilde{q}_{d,i}$, $\tilde{p}_{d,i}$ will converge.

$$\lim_{t\to\infty}\tilde{q}_{d,i}(t)=0,\ \lim_{t\to\infty}\tilde{p}_{d,i}(t)=0, \tag{8.31}$$

the proof is thus completed. □

Remark 8.2 The inequality $\beta > 2\pi c_4 + p_d^* + \beta^*$ in Eq. (8.24) ensures that all USVs continuously orbit the moving target vessel under the bearing-only protocol (8.20). This condition guarantees that $\bar{\sigma}_i(t)$, $i \in \mathcal{V}$, as given in (8.29), is persistently exciting, which is necessary for the convergence of the estimator (8.19) as shown in Lemma 8.4. This differs from conventional fencing strategies, where agents simply maintain fixed relative positions around the moving target (see, e.g., [17, 18, 26]).

Remark 8.3 Given that c_4 in (8.24) is a control gain designed for the controller (8.20) and $\beta^* \in \mathbb{R}^+$ is a predetermined constant, the parameter β can be set in advance to meet the condition (8.24), ensuring all parameters are feasible in practice. To mitigate the need for global information p_d^*, β can be estimated by using an overestimate $\widehat{p}_d^*$. Specifically, if each USV uses a value $\widehat{p}_d^* > p_d^*$, then β can be chosen to satisfy $\beta > 2\pi c_4 + \widehat{p}_d^* + \beta^*$, thus avoiding the requirement of knowing $p_d^* < \widehat{p}_d^*$. Moreover, the controller (8.20) can be implemented in a distributed manner because all necessary parameters for condition (8.24) can be set before executing the individual control laws.

Theorem 8.1 *Given Assumptions 8.1 through 8.2, the closed-loop system consisting of* (8.1), (8.3), (8.19), *and* (8.20) *successfully solves Problem 8.1, with the conditions* $\lim_{t\to\infty}\widetilde{u}_i(t)=0,\ \lim_{t\to\infty}\widetilde{v}_i(t)=0$ *being satisfied.*

Proof It follows from the controller (8.20) and the closed-loop system (8.17) that

$$\begin{aligned}\begin{bmatrix}\dot{\rho}_i\\ \rho_i\dot{\theta}_i\end{bmatrix} &= \begin{bmatrix}c_3(\rho_d-\widehat{\rho}_i)+\cos\theta_i\widehat{v}_x+\sin\theta_i\widehat{v}_y\\ c_4(\varsigma_i-\varsigma_{i^-})+\beta-\sin\theta_i\widehat{v}_x+\cos\theta_i\widehat{v}_y\end{bmatrix}+\begin{bmatrix}\bar{u}_i\\ \bar{v}_i\end{bmatrix}-\begin{bmatrix}\cos\theta_i & \sin\theta_i\\ -\sin\theta_i & \cos\theta_i\end{bmatrix}\begin{bmatrix}v_x^d\\ v_y^d\end{bmatrix}\\ &= \begin{bmatrix}c_3(\rho_d-\widehat{\rho}_i)+\eta_i\\ c_4(\varsigma_i-\varsigma_{i^-})+\beta+\omega_i\end{bmatrix}\end{aligned} \tag{8.32}$$

where

$$\begin{aligned}\eta_i &= \bar{u}_i+\cos\theta_i(\widehat{v}_x-v_x^d)+\sin\theta_i(\widehat{v}_y-v_y^d),\\ \omega_i &= \bar{v}_i-\sin\theta_i(\widehat{v}_x-v_x^d)+\cos\theta_i(\widehat{v}_y-v_y^d),\end{aligned}$$

which implies that the closed-loop system is decoupled to be

$$\begin{aligned}\dot{\rho}_i &= c_3(\rho_d-\widehat{\rho}_i)+\eta_i,\\ \dot{\theta}_i &= \frac{c_4(\varsigma_i-\varsigma_{i^-})+\beta}{\rho_i}+\frac{\omega_i}{\rho_i}.\end{aligned} \tag{8.33}$$

Based on the condition of $\lim_{t\to\infty}\widetilde{u}_i(t)=0,\ \lim_{t\to\infty}\widetilde{v}_i(t)=0$, it, together with (8.17) gives that

$$\lim_{t\to\infty}\bar{u}_i(t)=0,\ \lim_{t\to\infty}\bar{v}_i(t)=0. \tag{8.34}$$

Additionally, it follows from Lemma 8.4 that (8.3) that

$$\lim_{t\to\infty}\widehat{v}_x(t)-v_x^d=0,\ \lim_{t\to\infty}\widehat{v}_y(t)-v_y^d=0. \tag{8.35}$$

Then, according to Eqs. (8.34), (8.35) and η_i, ω_i in Eq. (8.32) together, one has that

$$\lim_{t\to\infty}\eta_i(t)=0,\ \lim_{t\to\infty}\omega_i(t)=0. \tag{8.36}$$

Define $\widetilde{\rho}_i := \rho_i-\rho_d$ to be the radius error, and take the temporal derivative of $\widetilde{\rho}_i$ along the dynamics of (8.33)

$$\dot{\widetilde{\rho}}_i = c_3(\rho_d-\rho_i)+c_3(\rho_i-\widehat{\rho}_i)+\eta_i = -c_3\widetilde{\rho}_i+\bar{\eta}_i \tag{8.37}$$

where $\bar{\eta}_i = c_3(\rho_i - \widehat{\rho}_i) + \eta_i$. Meanwhile, recalling the estimated $\widehat{q}_{d,i}$ and $\widehat{p}_d$ in Lemma 8.4, one has that

$$\lim_{t\to\infty} \widehat{\rho}_i(t) = \|\widehat{q}_{d,i}(t)\| = \rho_i(t), \tag{8.38}$$

which further follows from Eqs. (8.36) and (8.38) that $\lim_{t\to\infty} \bar{\eta}_i(t) = 0$. By formulating Eq. (8.33) in a compact form to be

$$\dot{\widetilde{\rho}} = -c_3\widetilde{\rho} + \bar{\eta}, \tag{8.39}$$

with $\widetilde{\rho} = [\widetilde{\rho}_1, \widetilde{\rho}_2, \ldots, \widetilde{\rho}_n]^{\mathrm{T}}$ and $\bar{\eta} = [\bar{\eta}_1, \bar{\eta}_2, \ldots, \bar{\eta}_n]^{\mathrm{T}} \in \mathbb{R}^n$. We select a Lyapunov function $V(\widetilde{\rho}) = \frac{1}{2}\widetilde{\rho}^{\mathrm{T}}\widetilde{\rho}$, satisfying

$$\dot{V}(\widetilde{\rho}) = -c_3\widetilde{\rho}^{\mathrm{T}}\widetilde{\rho} + \widetilde{\rho}^{\mathrm{T}}\bar{\eta} \le -(1-m_1)c_3\|\widetilde{\rho}\|^2 + \|\widetilde{\rho}\|(\|\bar{\eta}\| - m_1c_3\|\widetilde{\rho}\|) \tag{8.40}$$

with $m_1 \in (0, 1)$. In this way, it can be deduced that

$$\dot{V}(\widetilde{\rho}) \le -(1-m_1)c_3\|\widetilde{\rho}\|^2,$$

if $\|\widetilde{\rho}\| \ge \|\bar{\eta}\|/(m_1c_3)$. If $\|\widetilde{\rho}(0)\| \ge \|\bar{\eta}(0)\|/(m_1c_3)$, then $\dot{V}(\widetilde{\rho}) \le 0$, which indicates that $\|\widetilde{\rho}(t)\|$ approaches the ball of radius $\|\bar{\eta}\|/(m_1c_3)$. Otherwise, or, $\|\widetilde{\rho}(0)\| < \|\bar{\eta}(0)\|/(m_1c_3)$, it is clear that $\|\widetilde{\rho}(0)\|$ has already been in the ball of radius $\|\bar{\eta}\|/(m_1c_3)$. Both scenarios fulfill the definition of ISS [24], which implies that for any initial condition $\|\widetilde{\rho}(0)\|$, $\dot{V}(\widetilde{\rho}) \le 0$. Consequently, the closed-loop system (8.39) is ISS with respect to $|\bar{\eta}|$, satisfying the following condition

$$\|\widetilde{\rho}(t)\| \le \beta(\|\widetilde{\rho}(0\|, t)) + \gamma\big(\sup_{0\le\tau\le t} \|\bar{\eta}(\tau)\|\big), \forall t > 0. \tag{8.41}$$

Here, $\beta(\cdot)$ is a $\mathcal{KL}$ function with $\kappa_1 = 1/(m_1c_3)$ as the system gain (see [40]). Given that $\lim_{t\to\infty} \bar{\eta}_i(t) = 0$ and the $\mathcal{KL}$ function satisfies $\lim_{t\to\infty} \beta((\|\widetilde{\rho}\|, t)) = 0$, one has that $\lim_{t\to\infty} \widetilde{\rho}(t) = 0$ in Eq. (8.41).

Next, we will prove that the USVs are evenly distributed around the moving target vessel. Referring to the definition of ς_i in (8.7), it can be derived from (8.33) that

$$\begin{aligned}\dot{\varsigma}_i = \dot{\theta}_{i^+} - \dot{\theta}_i &= \frac{c_4(\varsigma_{i^+} - \varsigma_i) + \beta}{\rho_{i^+}} - \frac{c_4(\varsigma_i - \varsigma_{i^-}) + \beta}{\rho_i} + \frac{\omega_{i^+}}{\rho_{i^+}} - \frac{\omega_i}{\rho_i} \\ &= \frac{c_4(\varsigma_{i^+} - 2\varsigma_i + \varsigma_i)}{\rho_d} + \frac{c_4}{\rho_d}\epsilon_i + \varpi_i,\end{aligned} \tag{8.42}$$

where i^+, i^- denotes the pre-neighbor and next-neighbor of the USV i, respectively, and

$$\epsilon_i := \frac{(\varsigma_{i^+} - \varsigma_i)(\rho_{i^+} - \rho_d)}{\rho_{i^+}} - \frac{(\varsigma_i - \varsigma_{i^-})(\rho_i - \rho_d)}{\rho_i},$$
$$\varpi_i := \frac{\beta(\rho_{i^+} - \rho_i)}{\rho_i \rho_{i^+}} + \frac{\omega_{i^+}\rho_i - \omega_i \rho_{i^+}}{\rho_i \rho_{i^+}}. \tag{8.43}$$

Rewrite (8.42) in a compact form to be

$$\dot{\varsigma} = -\frac{c_4}{\rho_d} L(\mathcal{V})\varsigma + \frac{c_4}{\rho_d}\epsilon + \varpi, \tag{8.44}$$

where $\varsigma = [\varsigma_1, \varsigma_2, \ldots, \varsigma_n]^\mathrm{T}$, $\epsilon = [\epsilon_1, \epsilon_2, \ldots, \epsilon_n]^\mathrm{T}$, $\varpi = [\varpi_1, \varpi_2, \ldots, \varpi_n]^\mathrm{T}$ and $L(\mathcal{V}) \in \mathbb{R}^{n \times n}$ is a Laplacian matrix corresponding to the strongly connected graph $\mathcal{G}(\mathcal{V}, \mathcal{E})$.

Define $\tilde{\varsigma} = [\tilde{\varsigma}_1, \tilde{\varsigma}_2, \ldots, \tilde{\varsigma}_n]^\mathrm{T} \in \mathbb{R}^n$ to be the error vector between the separation angles and the desired angles as

$$\tilde{\varsigma} = \mathbf{1} \otimes \frac{2\pi}{n} - \varsigma \tag{8.45}$$

where a vector $\mathbf{1} \in \mathbb{R}^n$. Then, it follows from Eqs. (8.44) and (8.45) that

$$\dot{\tilde{\varsigma}} = \frac{c_4}{\rho_d} L(\mathcal{V})\varsigma + \frac{c_4}{\rho_d}\epsilon + \varpi = \frac{c_4}{\rho_d} L(\mathcal{V})\left(\mathbf{1} \otimes \frac{2\pi}{n} - \tilde{\varsigma}\right) + \frac{c_4}{\rho_d}\epsilon + \varpi. \tag{8.46}$$

Note that ϵ_i, ϖ_i is

$$\epsilon_i := \frac{(\tilde{\varsigma}_i - \tilde{\varsigma}_{i^+})(\rho_{i^+} - \rho_d)}{\rho_{i^+}} - \frac{(\tilde{\varsigma}_{i^-} - \tilde{\varsigma}_i)(\rho_i - \rho_d)}{\rho_i}. \tag{8.47}$$

From the fact that the matrix $L(\mathcal{V})$ satisfies $L(\mathcal{V})\mathbf{1} = \mathbf{0}$, the dynamic of $\dot{\tilde{\varsigma}}$ in (8.46) is

$$\dot{\tilde{\varsigma}} = -\frac{c_4}{\rho_d} L(\mathcal{V})\tilde{\varsigma} + \frac{c_4}{\rho_d}\epsilon + \varpi. \tag{8.48}$$

At the same time, it, together with Eq. (8.41), gives that

$$\rho_i(t) \geq \rho_d - \beta(\|\tilde{\rho}(0\|, t)) - \gamma\big(\sup_{0 \leq \tau \leq t} \|\bar{\eta}(\tau)\|\big). \tag{8.49}$$

Since $\beta(\|\tilde{\rho}(0\|, t)) > 0$, $\gamma\big(\sup \|\bar{\eta}(\tau)\|\big) > 0$ satisfies $\lim_{t\to\infty} \beta((\|\tilde{\rho}\|, t)) = 0$ and $\lim_{t\to\infty} \bar{\eta}_i(t) = 0$, one has that there exists $t_1 > 0$ such that $\rho_i(t) > \bar{\rho}, i \in \mathcal{V}, t > t_1$ with a constant $\bar{\rho} > 0$ and $\rho_d > 0$. Due to $\tilde{\varsigma}_i - \tilde{\varsigma}_{i^+} \leq 2\pi, i \in \mathcal{V}$, it can be deduced from Eqs. (8.43) and (8.47) that

$$\|\epsilon_i\| \leq \frac{2\pi|\rho_{i^+} - \rho_d|}{\bar{\rho}} + \frac{2\pi|\rho_i - \rho_d|}{\bar{\rho}}, \ \|\varpi_i\| \leq \frac{\beta|\rho_{i^+} - \rho_i|}{\bar{\rho}^2} + \frac{\omega_{i^+}}{\bar{\rho}} + \frac{\omega_i}{\bar{\rho}}, \tag{8.50}$$

which indicates that $\|\epsilon_i\|$, $\|\varpi_i\|$ are both bounded. Additionally, it can be deduced from Eqs. (8.36) and (8.41) that $\lim_{t\to\infty}\|\epsilon_i\| = 0$, $\lim_{t\to\infty}\|\varpi_i\| = 0$, which indicates that

$$\lim_{t\to\infty} \frac{c_4}{\rho_d}\|\epsilon(t)\| + \|\varpi(t)\| = \mathbf{0}, \tag{8.51}$$

where $\mathbf{0} \in \mathbb{R}^n$ is a zero vector. We consider a nominal system of (8.48) to be

$$\dot{\tilde{\varsigma}} = -\frac{c_4}{\rho_d} L(\mathcal{V})\tilde{\varsigma}. \tag{8.52}$$

Since $\mathcal{G}(\mathcal{V}, \mathcal{E})$ is time-changing during the target fencing mission, we select a Lyapunov candidate to be

$$V(\tilde{\varsigma}) = \frac{1}{2}\tilde{\varsigma}^{\mathsf{T}}\tilde{\varsigma}, \tag{8.53}$$

which is piecewise continuously differentiable. Given that the topology $\mathcal{G}(\mathcal{V}, \mathcal{E})$ remains unchanged, the corresponding $L(\mathcal{V})$ is symmetric and strongly connected. Taking the right upper Dini temporal derivative of $V(\tilde{\varsigma})$ (see [19]) along the trajectory described by (8.52), we obtain

$$D^+V(\tilde{\varsigma}) = \frac{1}{2}(\tilde{\varsigma}^{\mathsf{T}}\dot{\tilde{\varsigma}} + \dot{\tilde{\varsigma}}^{\mathsf{T}}\tilde{\varsigma}) = -\frac{c_4}{\rho_d}\tilde{\varsigma}^{\mathsf{T}}L(\mathcal{V})\tilde{\varsigma}^{\mathsf{T}}. \tag{8.54}$$

From the condition that $\mathbf{1}^{\mathsf{T}}\tilde{\varsigma} = 2\pi - \mathbf{1}^{\mathsf{T}}\varsigma = 2\pi - 2\pi = 0$, $\varsigma \neq \mathbf{0}$ and Eqs. (8.7) and (8.45), it, together with Lemma 8.2, gives that

$$\frac{dV(\tilde{\varsigma})}{dt} \leq -\frac{c_4}{\rho_d}\lambda_2(L(\mathcal{V}))\|\tilde{\varsigma}\|^2. \tag{8.55}$$

Due to the minimal-neighbor strategy outlined in Assumption 8.2, the graph $\mathcal{G}(\mathcal{V}, \mathcal{E})$ belongs to a finite set. Furthermore, there exists a lower bound $\chi > 0$ such that $\lambda_2(L(\mathcal{V})) > \chi$, which implies that

$$\frac{dV(\tilde{\varsigma})}{dt} \leq -\frac{c_4}{\rho_d}\chi\|\tilde{\varsigma}\|^2 \leq 0, \forall t > t_1. \tag{8.56}$$

Then, $\tilde{\varsigma}$ in the nominal system (8.52) satisfies $\lim_{t\to\infty}\tilde{\varsigma}_i(t) = 0$, exponentially.

According to Lemma 8.1 and (8.51), it follows that the perturbed system (8.48) satisfies $\lim_{t\to\infty}\tilde{\varsigma}(t) = 0$, which means $\lim_{t\to\infty}\varsigma_i(t) = 2\pi/n$ as stated in (8.8), resulting in an even phase distribution. Thus, the proof is completed. □

Remark 8.4 The proposed estimator (8.19) is more economical because it relies solely on bearing information about the moving target vessel, without needing to know the target's velocity. This is in contrast to the frameworks discussed in [33–35], where it is typically assumed that only a subset of agents can access the target's velocity. Bearing sensors are generally less expensive than velocity sensors, making the proposed approach less costly.

8.3.2 Vessel Regulation Control

After completing the target fencing mission for the multi-USV system with upper-level signals u_i^r and v_i^r, and ensuring that $\widetilde{u}_i, \widetilde{v}_i$ approach zero, we can proceed to address Problem 8.2. This involves designing the actuator inputs $\tau_{i,1}$, $\tau_{i,2}$ for the USV i, governed by the dynamic system (8.2), so that u_i and v_i converge to u_i^r and v_i^r, respectively.

According to $\widetilde{u}_i, \widetilde{v}_i$ in Eq. (8.16), one has that the actuator inputs $\tau_{i,1}$, $\tau_{i,2}$ for USV i are designed

$$\begin{aligned}\tau_{i,1} =& \frac{1}{k_3 u_i^r}(\dot{u}_i^r u_i - k_1 u_i^r u_i - k_2 u_i^r v_i r_i - c_1 u_i \widetilde{u}_i),\\ \tau_{i,2} =& \frac{1}{k_5}(-k_4 r_i - c_3 \widetilde{r}_i + \dot{r}_i^r),\end{aligned} \tag{8.57}$$

where $c_1, c_2, c_3 \in \mathbb{R}^+$ are positive parameters to be designed later, $\widetilde{r}_i$ and r_i^r the yaw velocity error and the desired yaw velocity, respectively, to be formulated as

$$\widetilde{r}_i = r_i - \widetilde{r}_i^r,\ r_i^r = \frac{1}{k_7 u_i}(-k_6 v_i + \dot{v}_i^r - c_2 \widetilde{v}_i). \tag{8.58}$$

Theorem 8.2 *A USV dynamic system, governed by* (8.2) *and* (8.57)*, will solve Problem 8.2 by ensuring* $\lim_{t\to\infty} \widetilde{u}i(t) = 0$ *and* $\lim t \to \infty \widetilde{v}_i(t) = 0$ *for the designed signals* u_i^r *and* v_i^r*, only if the initial condition* $u_i(0) > 0$ *is satisfied.*

Proof From the definition of $\widetilde{u}_i$ in (8.16) and the dynamic equation $\dot{u}_i$ in (8.2), one has that

$$\dot{\widetilde{u}}_i = k_1 u_i + k_2 v_i r_i + k_3 \tau_{i,1} - \dot{u}_i^r. \tag{8.59}$$

Then, it follows from Eqs. (8.57) and (8.59) that

$$\begin{aligned}\dot{\widetilde{u}}_i =& k_1 u_i + k_2 v_i r_i + k_3 \left(\frac{1}{k_3 u_i^r}(\dot{u}_i^r u_i - k_1 u_i^r u_i - k_2 u_i^r v_i r_i - c_1 u_i \widetilde{u}_i) \right) - \dot{u}_i^r,\\ =& \frac{\dot{u}_i^r u_i - \dot{u}_i^r u_i^r - c_1 u_i \widetilde{u}_i}{u_i^r}.\end{aligned} \tag{8.60}$$

Since the closed-loop system (8.60) is time-invariant, and according to the definitions of non-autonomous and autonomous systems given in [24], this indicates that (8.60) is an autonomous system. Then, we select a Lyapunov function candidate to be

$$V(u_i, u_i^r) = \left(\frac{u_i - u_i^r}{u_i}\right)^2, \tag{8.61}$$

whose temporal derivative along the trajectory of (8.60) is

$$\dot{V}(u_i, u_i^r) = \frac{2\widetilde{u}_i u_i^2(\dot{u}_i - \dot{u}_i^r) - 2\widetilde{u}_i^2 u_i \dot{u}_i}{u_i^4} = -\frac{2c_1\widetilde{u}_i^2}{u_i^2} \leq 0. \tag{8.62}$$

It follows that $\dot{V}(u_i, u_i^r) = 0$ only when $\widetilde{u}i = 0$. This implies that the largest invariant set in $\{u_i | \dot{V}(u_i, u_i^r) = 0\}$ consists solely of the point $\{u_i^r\}$, making this invariant set a compact (bounded and closed) set. According to LaSalle's invariance principle [24], the trajectory of u_i converges to u_i^r, i.e., $\lim_{t\to\infty} \widetilde{u}_i(t) = 0$.

Additionally, $V(u_i, u_i^r)$ is also bounded. The initial condition $u_i(0) > 0$ provided in Theorem 8.2 guarantees that $u_i(t) > 0$ for all future times. This will be utilized in the next proof to establish that $\lim_{t\to\infty} \widetilde{v}_i(t) = 0$.

Substituting the desired yaw velocity r_i^r in Eq. (8.58) into the dynamic $\dot{v}_i$ in (8.2) yields

$$\dot{\widetilde{v}}_i = k_6 v_i + k_7 u_i r_i^r + k_7 u_i \widetilde{r}_i - \dot{v}_i^r = -c_2\widetilde{v}_i + k_7 u_i \widetilde{r}_i. \tag{8.63}$$

From the dynamic equation $\dot{r}_i$ in (8.2), it can be deduced that the derivative of $\widetilde{r}_i$ in (8.58) is

$$\dot{\widetilde{r}}_i = k_4 r_i + k_5 \tau_{i,2} - \dot{r}_i^r. \tag{8.64}$$

Analogously, combining Eqs. (8.57) and (8.64) together yields

$$\dot{\widetilde{r}}_i = k_4 r_i + k_5\left(\frac{1}{k_5}(-k_4 r_i - c_3\widetilde{r}_i + \dot{r}_i^r)\right) - \dot{r}_i^r = -c_3\widetilde{r}_i, \tag{8.65}$$

which indicates that $\lim_{t\to\infty} \widetilde{r}_i(t) = 0$ exponentially.

Due to the upper-level signals u_i^r, v_i^r and the signal error $\widetilde{u}_i$ are all bounded, one has u_i is bounded, which implies that $\lim_{t\to\infty} k_7 u_i \widetilde{r}_i(t) = 0$. Considering the dynamic of $\widetilde{v}_i$ in Eq. (8.63), one can conclude that $\lim_{t\to\infty} \widetilde{v}_i(t) = 0$ exponentially, which completes the proof. □

Remark 8.5 The term $\tau_{i,1}$ in (8.57) not only facilitates convergence to u_i^r, but also ensures that $u_i(t) > 0$ as long as $u_i(0) > 0$. This is essential for maintaining the desired yaw velocity r_i^r as defined in Eq. (8.58), thereby providing an improvement over the existing work in [17].

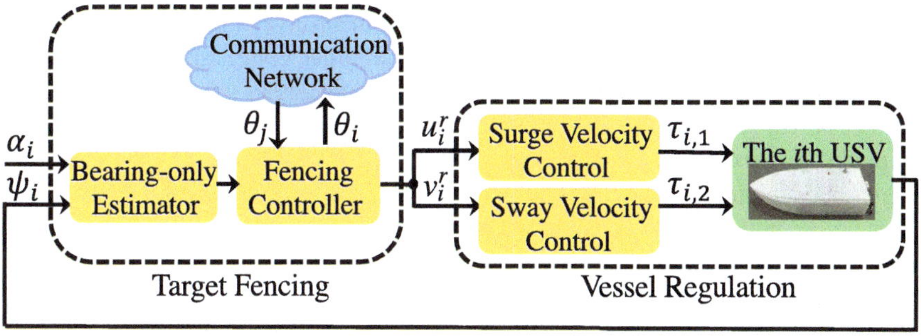

Fig. 8.3 Closed-loop structure of the bearing-only target-fencing controller for multi-USV systems

8.4 Algorithm Verification

In this section, we conduct bearing-only target-fencing experiments using our indoor multi-USV platform, which includes a motion capture system with eight Optitrack Flex 3 infrared cameras, a monitor station, three 300 mm-long 3D-printed HUSTER-0.3 USVs, a target vessel, and a pool measuring 4000 mm ×4000 mm ×500 mm, as shown in Fig. 8.4. Each USV is equipped with two 5V DC motors, two 150 mm transmission shafts, two Freescale Mini256ABC speed encoders, onboard 2.4 GHz wireless components (NRF24L01), three infrared emitters (peak wavelength: 850 nm), two LEDs, and an LI-PO battery (7.4 V, 1300 mAh, 25C). All these components are controlled by a microcontroller (STM32F103CBT6). For detailed information on the components and parameter settings, refer to [26].

The working principle of the platform is outlined as follows: First, the motion capture system detects the embedded infrared emitters on each USV using multiple infrared cameras, allowing it to calculate the positions and velocities of the USVs

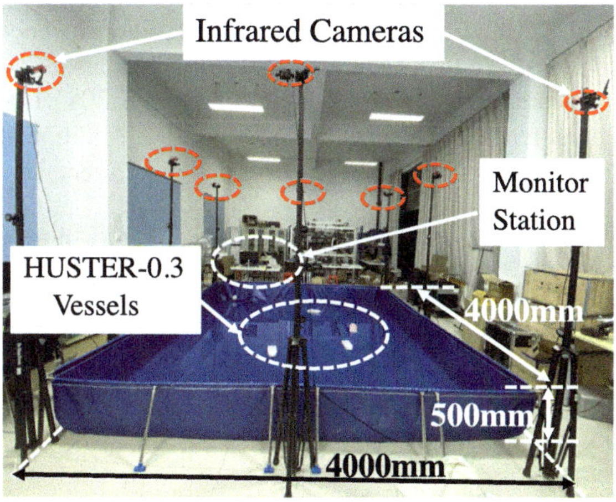

Fig. 8.4 The real map of the indoor multi-USV platform

with high accuracy-$\pm$3mm for position and $\pm$5 mm/s for velocity. Next, the monitor station calculates the bearing angles between the target vessel and the USVs based on the detected positions and velocities. These bearing angles are then transmitted to each USV via 2.4 GHz wireless components at a frequency of 100 Hz. Additionally, using the neighboring bearing angles as described in Assumption 8.2, the separation angles between pre- and next-neighbors are also communicated among the USVs via the wireless network. This enables each USV to estimate the target's velocity based on the neighboring bearing angles, allowing it to generate the individual control signals needed to perform the target fencing mission.

Precisely, the multi-USV system includes three USVs and a target vessel. We performed classical zigzag tracking experiments (refer to [17]) and collected data with a sampling interval of 0.1 s. Through least squares identification on this data, the dynamic parameters k_i for $i = 1, 2, \ldots, 7$ were estimated as follows: $k_1 = 0.9969$, $k_2 = -0.0731$, $k_3 = 0.00062$, $k_4 = 0.2491$, $k_5 = 0.00014$, $k_6 = 0.9877$ and $k_7 = 0.0081$. At the start of the experiment, the three USVs were randomly positioned within the pool area of $[-1000, 0]$ mm $\times$ $[-1500, 0]$ mm, with the total pool size being $[-1500, 2500]$ mm $\times$ $[-2000, 2000]$ mm. The initial velocities for the USVs were set as $u_i(0) = 10$ mm/s, $v_i(0) = 0$ mm/s, and $r_i(0) = 0$ rad/s for $i = 1, 2, 3$. The target vessel's position was randomly placed within $[0, 500]$ mm $\times$ $[0, 500]$ mm, and its velocity was set to $p_d = [40, 40]^\top$ mm/s, ensuring $\|p_d\| < p_d^* = 80$ mm/s according to Assumption 8.1. The desired radius was set to $\rho_d = 500$ mm. Following Theorem 8.1, we configured the estimator and controller parameters as $c_1 = 1$, $c_2 = 0.01$, $c_3 = 1$, and $c_4 = 1$. Given these settings and the desired radius, we selected $\beta = 200$ for (8.20), which satisfies the condition in (8.24) with $\beta^* = 20$. The actuator input parameters for (8.57) were chosen as $c_5 = 2$, $c_6 = 1.2$, and $c_7 = 2$, in accordance with Theorem 8.2.

Based on the fencing behavior described in Theorem 8.1, Fig. 8.5 demonstrates the fencing trajectories of three USVs around a moving target vessel, which illustrates that all USVs effectively fence and rotate evenly around the target using only bearing measurements. Figure 8.6 shows the convergence of the estimated relative position errors $\widetilde{q}_{d,i}, i = 1, 2, 3$, and the estimated relative velocity errors $\widetilde{p}_{d,i}, i = 1, 2, 3$, between the USVs and the target vessel. Figure 8.7 indicates that the relative distances $\rho_i, i = 1, 2, 3$, converge to the desired radius $\rho_d = 500$ mm, with the separation angles approaching 120° within less than 30 s. This demonstrates the successful target fencing as defined in Definition 8.1. Additionally, Fig. 8.8 captures a snapshot of the steady-state target fencing at the 24th second, as shown in Fig. 8.5.

Remark 8.6 By applying least square identification to zigzag tracking data with persistently exciting (PE) input signals, the vehicle parameter errors remain generally bounded. These errors are effectively managed by the vessel regulation law (8.57). The framework depicted in Fig. 8.3 integrates a high-level multi-vessel target fencing control with a low-level single-vessel regulation. This hierarchical closed-loop system benefits from robustness against bounded identification errors, as guaranteed by Theorem 8.1 and grounded in ISS theory. Consequently, experimental results

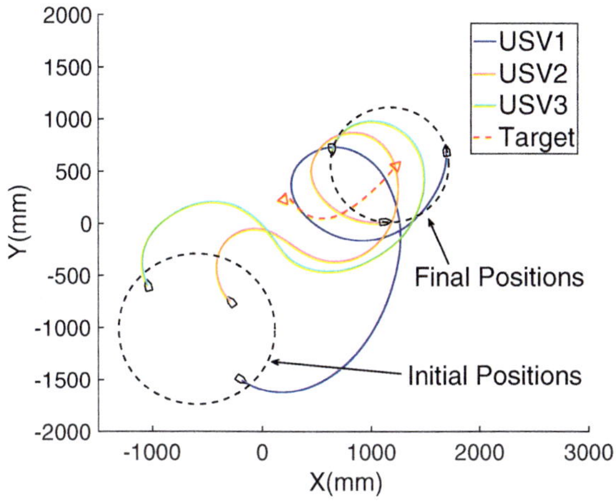

Fig. 8.5 A bearing-only target fencing process of a motional target vessel with three USVs, where the colored lines are the moving trajectories of the three USVs whereas the red dashed line is the target vessel's trajectory

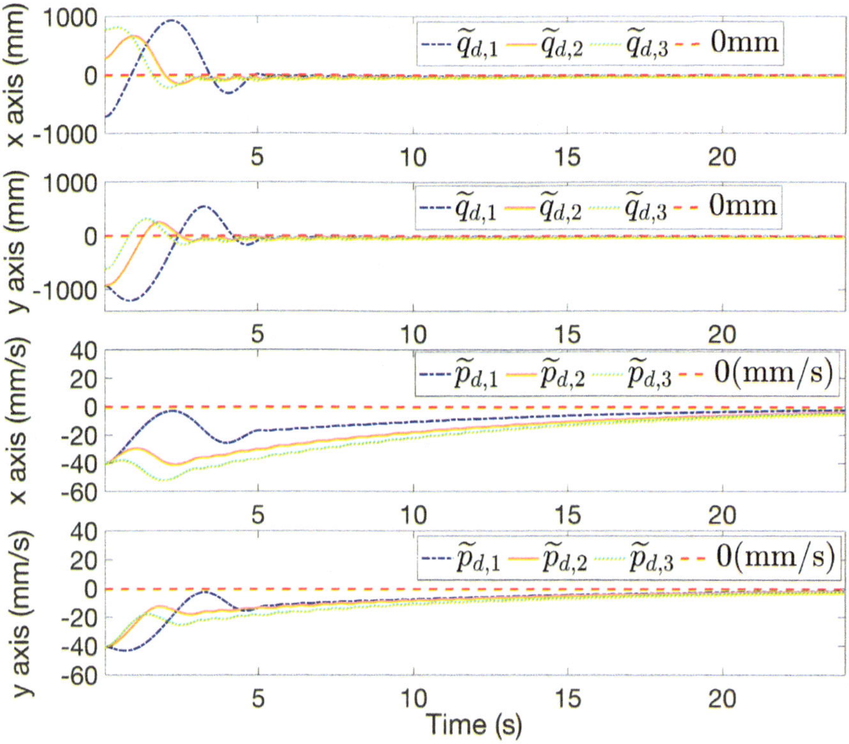

Fig. 8.6 Temporal evolution of the estimated relative-position errors $\widetilde{q}_{d,i}$, $i = 1, 2, 3$, and the estimated relative-velocity errors $\widetilde{p}_{d,i}$, $i = 1, 2, 3$, between the USVs and the target vessel

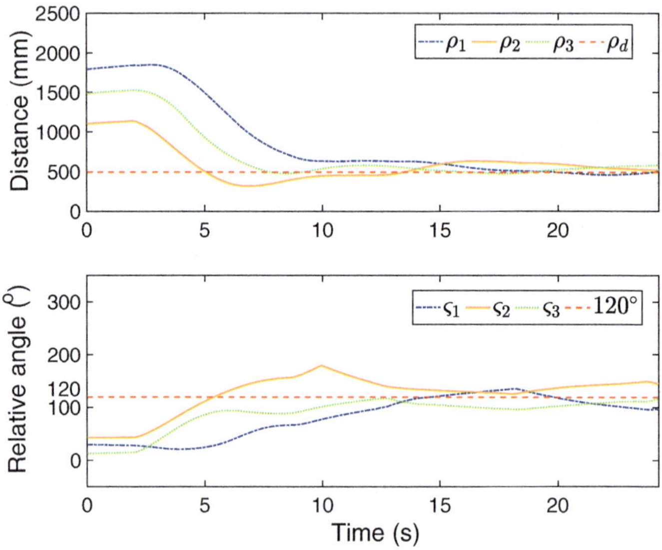

Fig. 8.7 Temporal evolution of the relative distances $\rho_i, i = 1, 2, 3$, and the separation angles $\varsigma_i, i = 1, 2, 3$

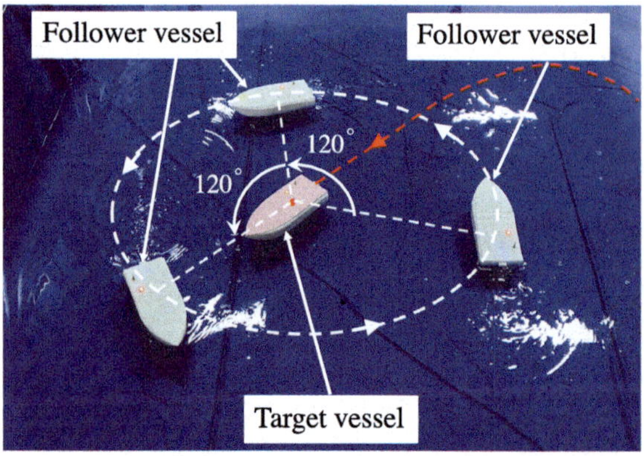

Fig. 8.8 A snapshot of the target-fencing mission with a motional target vessel at 24th second from the start

demonstrate robustness against such errors (see Figs. 8.5, 8.6, 8.7 and 8.8). To further enhance the multi-USV system's robustness against larger identification errors, one could incorporate terms for coefficient uncertainties and external noise into the vessel dynamics model (8.2). Seeking robust formation controllers, as discussed in recent developments (e.g., [39]), may provide additional support.

Remark 8.7 Note that the indoor multi-USV experimental platform incorporates a collision avoidance mechanism. When the distance between any two USVs falls below a specified safe threshold, the involved USV will enter a safe mode: the rear USV will either slow down, deviate, or stop to avoid a collision until the distance exceeds the safe threshold again. After the safe distance is restored, the system will revert to the original distributed fencing control laws (8.20) and (8.57). Collision avoidance can be considered a disturbance to the closed-loop system, which makes theoretical proof of motional target fencing convergence under these switching conditions more challenging. Nevertheless, in a sufficiently large pool, each USV will have enough space to perform collision avoidance maneuvers effectively. As the USVs gradually form a near-circular arrangement, they will remain sufficiently separated, reducing the need for further collision avoidance. Consequently, the proposed controllers (8.20) and (8.57) demonstrate robustness against disturbances caused by collision avoidance, thereby ensuring the effectiveness of motional target fencing.

8.5 Conclusion

In this chapter, we have developed and demonstrated a distributed bearing-only control strategy for networked USVs to achieve target fencing around a moving vessel. The proposed method has offered the advantage of calculating desired surge and sway velocities at the upper level, enabling the USVs to rotate uniformly around the moving target based solely on bearing measurements. Additionally, a single-vessel regulation has been employed to track these upper-level commands. The effectiveness of this control approach has been validated through experiments on an actual multi-USV platform. Future work will focus on enhancing the multi-USV fencing controller to handle challenges such as intermittent bearing measurements, communication delays, significant identification errors, strong external disturbances (like waves and tides), and the management of multiple moving targets.

References

1. E. Lefeber, K.Y. Pettersen, H. Nijmeijer, Tracking control of an underactuated ship. IEEE Trans. Control Syst. Technol. **11**(1), 52–61 (2003)
2. H. Ashrafiuon, K.R. Muske, L.C. McNinch, R.A. Soltan, Sliding-mode tracking control of surface vessels. IEEE Trans. Ind. Electron. **55**(11), 4004–4012 (2008)
3. K. Shojaei, Neural adaptive robust control of underactuated marine surface vehicles with input saturation. Appl. Ocean Res. **53**, 267–278 (2015)
4. W. He, Z. Yin, C. Sun, Adaptive neural network control of a marine vessel with constraints using the asymmetric barrier Lyapunov function. IEEE Trans. Cybern. **47**(7), 1641–1651 (2017)
5. N. Wang, H. He, Dynamics-level finite-time fuzzy monocular visual servo of an unmanned surface vehicle. IEEE Trans. Ind. Electron. **67**(11), 9648–9658 (2019)
6. Z. Peng, J. Wang, Q.-L. Han, Path-following control of autonomous underwater vehicles subject to velocity and input constraints via neurodynamic optimization. IEEE Trans. Ind. Electron. **66**(11), 8724–8732 (2018)
7. B. Liu, H.-T. Zhang, H. Meng, D. Fu, H. Su, Scanning-chain formation control for multiple unmanned surface vessels to pass through water channels. IEEE Trans. Cybern. **52**(3), 1850–1861 (2020)

8. N. Wang, G. Xie, X. Pan, S.-F. Su, Full-state regulation control of asymmetric underactuated surface vehicles. IEEE Trans. Ind. Electron. **66**(11), 8741–8750 (2019)
9. Z. Peng, J. Wang, D. Wang, Q.-L. Han, An overview of recent advances in coordinated control of multiple autonomous surface vehicles. IEEE Trans. Ind. Inform. **17**(2), 732–745 (2020)
10. L. Liu, D. Wang, Z. Peng, T. Li, C.L.P. Chen, Cooperative path following ring-networked under-actuated autonomous surface vehicles: algorithms and experimental results. IEEE Trans. Cybern. **50**(4), 1519–1529 (2020)
11. D. Li, G. Ma, W. He, S.S. Ge, T.H. Lee, Cooperative circumnavigation control of networked microsatellites. IEEE Trans. Cybern. **50**(10), 4550–4555 (2019)
12. M. Hassan, E. Aljuwaiser, R. Badr, A new on-line observer-based controller for leader-follower formation of multiple nonholonomic mobile robots. J. Franklin Inst. **355**(5), 2436–2472 (2018)
13. K. Shojaei, Observer-based neural adaptive formation control of autonomous surface vessels with limited torque. Robot. Auton. Syst. **78**, 83–96 (2016)
14. Z. Sun, G. Zhang, Y. Lu, W. Zhang, Leader-follower formation control of underactuated surface vehicles based on sliding mode control and parameter estimation. ISA Trans. **72**, 15–24 (2018)
15. F. Chen, W. Ren, Y. Cao, Surrounding control in cooperative agent networks. Syst. Control Lett. **59**(11), 704–712 (2010)
16. Y. Lan, G. Yan, Z. Lin, Distributed control of cooperative target enclosing based on reachability and invariance analysis. Syst. Control Lett. **59**(7), 381–389 (2010)
17. B. Liu, Z. Chen, H.-T. Zhang, X. Wang, T. Geng, H. Su, J. Zhao, Collective dynamics and control for multiple unmanned surface vessels. IEEE Trans. Control Syst. Technol. **28**(6), 2540–2547 (2019)
18. Z. Chen, A cooperative target-fencing protocol of multiple vehicles. Automatica **107**, 591–594 (2019)
19. J. Guo, G. Yan, Z. Lin, Local control strategy for moving-target-enclosing under dynamically changing network topology. Syst. Control Lett. **59**(10), 654–661 (2010)
20. X. Yu, L. Liu, Distributed circular formation control of ring-networked nonholonomic vehicles. Automatica **68**, 92–99 (2016)
21. X. Yu, L. Liu, Cooperative control for moving-target circular formation of nonholonomic vehicles. IEEE Trans. Autom. Control **62**(7), 3448–3454 (2016)
22. S. Shoja, M. Baradarannia, F. Hashemzadeh, M. Badamchizadeh, P. Bagheri, Surrounding control of nonlinear multi-agent systems with non-identical agents. ISA Trans. **70**, 219–227 (2017)
23. Y.J. Shi, R. Li, K.L. Teo, Cooperative enclosing control for multiple moving targets by a group of agents. Int. J. Control **88**(1), 80–89 (2015)
24. K.H. Khalil, *Nonlinear Systems* (Prentice Hall, Upper Saddle River, 2002)
25. C. Li, L. Chen, Y. Guo, Y. Lyu, Cooperative surrounding control with collision avoidance for networked Lagrangian systems. J. Franklin Inst. **355**(12), 5182–5202 (2018)
26. B.-B. Hu, H.-T. Zhang, J. Wang, Multiple-target surrounding and collision avoidance with second-order nonlinear multiagent systems. IEEE Trans. Ind. Electron. **68**(8), 7454–7463 (2020)
27. Z. Peng, Y. Jiang, J. Wang, Event-triggered dynamic surface control of an underactuated autonomous surface vehicle for target enclosing. IEEE Trans. Ind. Electron. **68**(4), 3402–3412 (2020)
28. X. Jin, Fault tolerant finite-time leader-follower formation control for autonomous surface vessels with LOS range and angle constraints. Automatica **68**, 228–236 (2016)
29. B.-B. Hu, H.-T. Zhang, B. Liu, H. Meng, G. Chen, Distributed surrounding control of multiple unmanned surface vessels with varying interconnection topologies. IEEE Trans. Control Syst. Technol. **30**(1), 400–407 (2021)
30. L. Kou, Z. Chen, J. Xiang, Cooperative fencing control of multiple vehicles for a moving target with an unknown velocity. IEEE Trans. Autom. Control **67**(2), 1008–1015 (2021)
31. R. Zheng, Y. Liu, D. Sun, Enclosing a target by nonholonomic mobile robots with bearing-only measurements. Automatica **53**, 400–407 (2015)

32. R. Li, Y. Shi, Y. Song, Localization and circumnavigation of multiple agents along an unknown target based on bearing-only measurement: A three dimensional solution. Automatica **94**, 18–25 (2018)
33. M. Deghat, I. Shames, B.D. Anderson, C. Yu, Localization and circumnavigation of a slowly moving target using bearing measurements. IEEE Trans. Autom. Control **59**(8), 2182–2188 (2014)
34. Z. Yang, S. Zhu, C. Chen, G. Feng, X. Guan, Entrapping a target in an arbitrarily shaped orbit by a single robot using bearing measurements. Automatica **113**, 108805 (2020)
35. L. Dou, C. Song, X. Wang, L. Liu, G. Feng, Target localization and enclosing control for networked mobile agents with bearing measurements. Automatica **118**, 109022 (2020)
36. T.I. Fossen, *Handbook of Marine Craft Hydrodynamics and Motion Control.* (Wiley, 2011)
37. R. Olfati-Saber, R.M. Murray, Consensus problems in networks of agents with switching topology and time-delays. IEEE Trans. Autom. Control **49**(9), 1520–1533 (2004)
38. S. Sastry, M. Bodson, *Adaptive Control: Stability* (Courier Corporation, Convergence and Robustness, 2011)
39. B.S. Park, S.J. Yoo, Connectivity-maintaining and collision-avoiding performance function approach for robust leader-follower formation control of multiple uncertain underactuated surface vessels. Automatica **127**, 109501 (2021)
40. A. Isidori, *Nonlinear Control Systems II* (Springer Science & Business Media, UK, 2012)

Chapter 9
Summary and Future Work

In this chapter, we summarize the whole monograph and further discuss future research topics of coordinated fencing missions.

9.1 Summary

This monograph has deeply explored the cooperative multi-robot fencing algorithm design and experiments. Compared with previous multi-robot fencing formation algorithms, the distinctive features of this monograph can be summarized as follows.

1. Firstly, most of the existing algorithms still focus on fixed-ordering fencing under fixed topologies. In contrast, this monograph delves deeper into convex-hull fencing, theoretically exploring more complex time-varying topology fencing, and achieving the flexible-ordering fencing formation.
2. Secondly, most of the existing fencing works remain at simple linear systems and typically consider controllers based on position feedback. This monograph integrates the complex USV dynamics, taking into account multiple factors such as bearing constraints and rotating formations, which thus brings it closer to practical applications.
3. Thirdly, most of the fencing works often remain in the stage of simulation validation. This monograph considers the cooperative fencing experiments using the self-established multi-USV platform, which further validates the effectiveness of the proposed algorithms, and thereby makes the conclusions more reliable and practically valuable.

B.-B. Hu and H.-T. Zhang, *Multi-Robot Systems*, Cognitive Intelligence and Robotics,
https://doi.org/10.1007/978-981-97-9214-6_9

Despite this monograph investigating a series of studies on the theory and application of cooperative fencing control, it still has limitations.

1. In Chap. 2, a distributed controller for the basic fencing problem of constant-speed targets is proposed, but it is limited to a constant-speed target.
2. In Chap. 3, we only focus on complex cooperative fencing control, primarily considering the uniform rotating formations and intermittent topology disturbances, but the target remains a constant-speed target as well, without addressing the challenging scenarios with a varying-velocity target.
3. In Chap. 4, we further explore the complex fencing of a varying-speed target, but the target dynamics are limited to some extent known, without considering the cooperative fencing problem where the target dynamics are completely unknown.
4. In Chap. 5, we propose the formal method for cooperative fencing of multi-robot systems. However, due to the CBF functions requiring smooth boundaries of functions, one has that such kinds of methods cannot apply to the polygon obstacle environments, which requires further investigations.
5. In Chap. 6, we leverage the adaptive controller for fencing multiple targets with different speeds, but only consider bounded speed variations, and collision avoidance between robots. It does not address the multiple targets with randomly varying speeds and collision avoidance between robots and multiple targets.
6. In Chap. 7, we consider the flexible fencing and subgrouping for multiple scattered targets. However, due to the introduction of mixed-integer quadratic programming (MIQP), such kind of methods require a high computation burden, which cannot be extended to a large-scale scenario. Additionally, it also cannot be used in polygon-obstacle environments as well.
7. In Chap. 8, we propose a cooperative fencing algorithm relying solely on bearing information. However, it only simulates bearing using multi-camera positioning systems and differential GPS calculations, without using monocular cameras to directly obtain bearing information. Meanwhile, the current work only considers a single type of sensor, without involving heterogeneous sensors.

In general, Chaps. 2–8 focus on distributed cooperative fencing control in open environments, achieving good simulation and experimental results. However, they have not been combined with more complex fencing tasks and dynamically changing narrow environments, including moving obstacles, and irrelevant targets. The limitations mentioned above will be further addressed and improved in future work.

9.2 Future Work

The research on cooperative fencing of multi-robot systems has been a hot topic in the robotics community. Although this monograph has conducted some research on the design and practical application of distributed cooperative fencing algorithms, it is still in the initial stage. There are still various unresolved issues that require further investigation in future work.

1. **(Heterogeneous Coordination)** Currently, most of the works on multi-robot cooperative fencing only consider the fencing tasks of homogeneous robots. These algorithms focus on how to drive a single type of robot to complete the cooperative fencing of a target, which is often only suitable for simple task scenarios. In some real-world scenarios, multiple types of robots with different features may be involved, which is necessary to fully utilize the different advantages of each type of robot and coordinate the impacts of differences among them to complete the corresponding fencing tasks. Therefore, one of the future directions can consider the cooperative fencing of heterogeneous robots, which may include multi-task allocation and real-time scheduling. Additionally, conducting multiple tasks in parallel according to actual scenarios is another interesting topic.
2. **(Cross-domain Coordination)** Currently, most of the works on multi-robot cooperative fencing only focus on designing 2D fencing missions for USVs or UGVs, which results in these algorithms being suitable for some unobstructed environments. However, in real environments, in addition to USVs and UGVs, there may also be a requirement for the coordination of UAVs, involving cooperative fencing control for the target in the air. Furthermore, other constraints such as obstacles, ships, buoys, channels, and reefs can significantly impact the control signals, making the fencing mission difficult to achieve. Therefore, one of the future directions can consider distributed cross-domain planning and decision-making of multi-robot systems, which is more challenging to design the distributed real-time planning of each robot's desired path and control signals in dynamic environments. It can enhance the practical application potential of cooperative operations with unmanned vehicles.
3. **(Game-Theoretic Fencing)** Currently, most of the works on multi-robot cooperative fencing only assume individual homogeneous intelligence, where the target-fencing mission is always achieved through collective motion of robots. Therein, the target's position usually is available and the target behaves cooperatively as well, typically being passively fenced without the awareness to actively escape. In reality, due to different performances among individual robots, they may exhibit individual intelligence while completing the overall fencing mission, aiming to minimize their costs as much as possible. The target also possesses intelligence and is capable of actively evading the fencing. One of the future directions can consider relevant game theory, focusing on the competitive game situation among individual robots during the fencing process, as well as the pursuit–escape game between the individuals and the target, making cooperative fencing intelligent. Moreover, due to the introduction of the game theory, there are some theoretical directions such as Nash equilibrium searching and convergences analysis, which are also very interesting.

4. **(Formal Feasibility Analysis)** Currently, most of the works on multi-robot cooperative fencing mainly focus on how to achieve the fencing task as a cooperative control objective, without considering the problem feasibility. In other words, there are few works considering whether the target-fencing mission can be achieved or not. Specifically, when facing a complex fencing problem, one of the future directions can consider how to quickly determine whether a feasible solution exists, i.e., finding sufficient and necessary conditions for individuals to achieve cooperative fencing.

The manufacturer's authorised representative in the EU is Springer Nature Customer Service Centre GmbH, Europaplatz 3, 69115 Heidelberg, Germany. If you have any concerns regarding our products, please contact ProductSafety@springernature.com

Printed and bound by CPI Group (UK) Ltd, Croydon, CR0 4YY

01/10/2026

02225919-0002